SpringerBriefs in Molecular Science

More information about this series at http://www.springer.com/series/8898

Rohit Srivastava · Narendra Yadav
Jayeeta Chattopadhyay

Growth and Form of Self-organized Branched Crystal Pattern in Nonlinear Chemical System

Springer

Rohit Srivastava
Department of Chemistry
Birla Institute of Technology, Mesra,
  Deoghar Off-Campus
Deoghar, Jharkhand
India

and

Department of Inorganic and Physical
  Chemistry
Indian Institute of Science
Bangalore, Karnataka
India

Narendra Yadav
Department of Space Engineering and
  Rocketry
Birla Institute of Technology, Mesra
Ranchi, Jharkhand
India

Jayeeta Chattopadhyay
Department of Chemistry
Birla Institute of Technology, Mesra,
  Deoghar Off-Campus
Deoghar, Jharkhand
India

ISSN 2191-5407   ISSN 2191-5415 (electronic)
SpringerBriefs in Molecular Science
ISBN 978-981-10-0863-4   ISBN 978-981-10-0864-1 (eBook)
DOI 10.1007/978-981-10-0864-1

Library of Congress Control Number: 2016935982

Printed on acid-free paper

This Springer imprint is published by Springer Nature
The registered company is Springer Science+Business Media Singapore Pte Ltd.

Preface

This book is aimed at all those who are interested to understand the self-organization and growth of branched crystal pattern mechanism in Belousov–Zhabotinsky (BZ) type oscillatory chemical reactions including all aspects relevant for physical chemists, chemical engineers, and materials scientists. The oscillatory BZ reaction is one of the systems which has laid a solid foundation in understanding many natural phenomena through the reaction–diffusion mechanism. This reaction has been used to study various nonlinear phenomena such as oscillation, excitability, chaos, spatiotemporal patterns, and propagation of chemical waves in different reaction conditions. The mechanism of pattern formation and their applications in diverse context of real systems are the key issues for extensive study of the oscillatory BZ reaction. Some common chemicals of BZ reaction are essential for designing a new or modified oscillatory chemical system. This allows estimating the possible products and molecular mechanisms behind the formation of complex spatiotemporal patterns with the help of some kinetic data and rate equations of classic BZ reaction. Thus, the BZ reaction system has become a model chemical system for the study of many physical, chemical, and biological processes of non-equilibrium states. For its esthetic appeal and richness of dynamic behaviors, the study of oscillatory reactions has advanced rapidly in recent years. Scientists have made efforts to seek insight into the attractive space–time structures that occur in interdisciplinary research fields, such as materials processing, electro-polymerization, self-assembly process, and designing of self-oscillating gels. The non-equilibrium crystallization phenomenon and growth of nanostructured crystal patterns are the two progressive areas where the role of chemical oscillations has been apparently exemplified. In the present investigation, the growth of some highly ordered crystal patterns have been studied under the BZ reaction conditions. Self-organization, reaction–diffusion, and chemical compositions have been quantified as elementary controlling factors where crystal growths, transformation of crystal textures, phase ordering, and the transition in crystal morphologies have been undertaken. Optimization of these crystal patterns has been carried out with different concentrations of BZ reagents, reaction temperatures, and the various analytical techniques.

Acknowledgments

First, we would like to thank everyone who will read this book. Our heartiest thanks to all our colleagues who worked tremendously for this work to make it successful. Their enormous effort and devotion motivated us to work harder. Our special thanks to the publisher, Springer (India) Pvt. Ltd. for providing us such great opportunity to share our knowledge with the entire scientific community. The greatest thanks go to our parents and beloved family members, without their support this book could have not been possible.

Acknowledgements

Contents

About the Authors

Dr. Rohit Srivastava was born in Pratapgarh in 1986. He got his B.Sc. and M.Sc. degrees from Dr. Ram Manohar Lohia Avadh University, Faizabad, India in 2005 and 2008, respectively. He received Master of Philosophy (M.Phil.) in Chemistry from Dr. Bhim Rao Ambedkar University Agra, India in 2010 and after that he worked as a project fellow on UGC major research project at Motilal Nehru National Institute of Technology, Allahabad, India. He obtained his Ph.D. from Birla Institute of Technology, Mesra, Ranchi, India, in 2015, where he investigated "Self-organization and growth of nanostructured branched crystal pattern in Belousov-Zhabotinsky type chemical reactions". He is currently working as postdoctoral fellow in Department of Inorganic and Physical Chemistry, Indian Institute of Science, Bangalore. He has published more than 15 research papers in high-impact international journal, one book chapter and one book. He has received awards like Dr. DS Kothari Post-doc fellowship of UGC, Government of India, Institute Post-doctoral Fellowship of IIT Bombay, and Institute fellowship (JRF) of BIT Mesra, Ranchi, India. He is reviewer of more than 5 international journals. His current research areas include polymer-based functional nanomaterials for biomedical applications, self-assembly, oscillatory chemical reactions, pattern formation in reaction–diffusion system, and electrospun-based nanofiber for biosensor applications.

Dr. Narendra Yadav received Ph.D. Degree in the Chemical Science from Birla Institute of Technology (BIT), Mesra, Ranchi, in 2013. Presently, he is working as Research Scientist in the Department of Space Engineering and Rocketry, BIT Mesra, since July 2010. He has been involved in research work for the growth of novel crystal patterns mediated with non-equilibrium crystallization and oscillatory chemical reactions. Dr. Yadav is also associated with various R&D activities in the area of high-energy materials (HEMs) for aerospace application. He has 10 papers in peer-reviewed journals and six publications in seminar and conference proceedings in general chemistry and propellant technology.

Dr. Jayeeta Chattopadhyay was born in Kolkata in 1981. She received her Bachelor of Science with Honours in Chemistry from Bethune College, Calcutta University. She got her Master of Science in Chemistry from Devi Ahilya University, Indore (2003) and Master of Technology in Fuels and Combustion from Birla Institute of Technology, Mesra, Ranchi (2005). She obtained her Ph.D. in New Energy Engineering with best doctoral thesis award from Seoul National University of Science and Technology, S. Korea (2010). She has obtained the prestigious Fast Track Young Scientist award from Department of Science and Technology, Government of India (2010). She has published more than 20 research and review articles in high-impact peer-reviewed international journals, and she is the owner of one international patent. She is the reviewer of more than seven high-impact international journals. Her research interest includes nanostructured materials for energy applications, and thermo-degradation of solid waste materials. She also works on oscillatory chemical reactions and pattern formation in reaction–diffusion system.

Abbreviations

AA	Acetyl acetone
ADA	Adipic acid
ADAA	Adipic anhydride
AOT	Water-in-oil bis(2-ethylhexyl) sulfosuccinate
ARD	Advection–reaction–diffusion
BA	Butanoic acid
BZ	Belousov–Zhabotinsky
CCD	Charge coupled device
CHD	1,4-Cyclohexanedione
DLA	Diffusion-limited-aggregation
DNA	Deoxy ribonucleic acid
EAA	Ethyl acetoacetate
EDS	Energy dispersive spectroscopy
EOR	Enhanced oil recovery
$[Fe(phen)_3]^{2+}$	Ferroin
FKN	Field–Körös–Noyes
FTIR	Fourier transform infrared
FWHM	Full width at high maxima
GFN	Growth front nucleation
SEM	Scanning electron microscopy
TEM	Transmission electron microscopy

Abstract

Self-organization phenomenon and various types of pattern formation, involving diffusion-driven mechanism might be observed in biological, chemical, and geo-chemical systems. Exciting phenomena such as precipitation, electric potential oscillations, chaos, fractal growth, etc., which are far from equilibrium, have captivated considerable attention in recent years. The spontaneous formation of nano-scale patterns seems to be a significant way to control the structure and morphology of various functional materials. Self-organization, based on interplay between reactions and diffusion, has been found to occur in a range of physical and chemical systems. The recent development of non-equilibrium crystallization phenomenon enables one to form spontaneous, coherent, and periodic patterns which are accompanied by molecular interactions. Among the different nanostructures, the dendrite (DLA and spherulitic) crystal patterns attract the attention of scientific community due to their importance in connection to some fractal growth phenomena and crystallography research. The growth of dendrite crystals is also an example which mimics several pattern-forming phenomena encountered in nature and biology. In this connection, the BZ reaction is one of the best prototype systems for exploring such self-organization based phenomenon.

The most widely studied oscillatory reaction is the BZ reaction that became model for investigating a wide range of intriguing pattern formations in chemical systems. So many modifications in classic version of BZ reaction have been carried out under various experimental conditions that demonstrate rich varieties of temporal oscillations and spatiotemporal patterns in non-equilibrium conditions. Mixed-mode versions of BZ reactions, which comprise a pair of organic substrates or dual metal catalysts, have displayed very complex oscillating behaviors and novel space–time patterns during reaction processes. These characteristic spatiotemporal properties of BZ reactions have attracted increasing attention of the scientific community in recent years because of its comparable periodic structures in electrochemical systems, polymerization process, and non-equilibrium crystallization phenomena. Instead non-equilibrium crystallization phenomenon which leads to the development of novel crystal morphologies in constraint of thermodynamic equilibrium conditions have been investigated and are said to be stationary periodic structures. Continuous efforts have been taken to analyze insightful mechanisms and roles of reaction-diffusion mechanism and self-organization in the growth of such periodic crystal patterns. In this book, non-equilibrium crystallization phenomena, leading to growth of some novel crystal patterns in dual organic substrate modes of oscillatory BZ reactions, have been discussed. Efforts have been made to find the experimental parameters at which transitions of the spherulitic crystal patterns take place.

Keywords Crystal growth · Surfaces · Diffusion · Chemical synthesis · Nanostructures

Chapter 1
Introduction

Experimental approaches to understand the physical and chemical phenomena facilitate a method by which scientists can handle many of the existing systems and manipulating them purposefully; this also makes possible to design new systems with tailor-made characteristics. In the recent years, many studies on self-organization in chemical reaction systems are carried out to assemble the experimentation facts. This shows how can simple mechanism of the reactions on molecular interactions are competent to form a number of attractive spatiotemporal structures by self-organization. These patterns were highly sensitive toward change in the concentrations and some other reaction parameters. On the other hand, the engineering of these systems is somehow different as suggested for chemically reacting systems. A rigid control which applied for the self-organizing process can significantly interfere to the fine interactions among molecular states of the system. The spontaneous activity of those systems can be steered in a desired direction by imposing feedback mechanisms and also by application of the controlled impulses. Thus, the transition between different states of organizations is initiated and a new form of system of collective behavior is achieved.

After the discovery of the chemical oscillation, the oscillatory Belousov–Zhabotinsky (BZ) reaction becomes model chemical system which displays oscillations, various periodic or spontaneous time, or spatial structural patterns in the contribution of self-organization and reaction–diffusion mechanism. In this regard, the nonlinear chemical dynamics emerged as an important methodological tool for the study of such complex patterning behaviors in interdisciplinary research fields. The basics of BZ self-organization are now apply all branches of material sciences, mathematics, biology, and engineering technology. However, the phenomenology, methodologies, mechanistic considerations, and theoretical aspects in diverse disciplines of science and technologies are partly or fully different for the studies of self-organizations.

© The Author(s) 2016

R. Srivastava et al., *Growth and Form of Self-organized Branched Crystal Pattern in Nonlinear Chemical System*, SpringerBriefs in Molecular Science, DOI 10.1007/978-981-10-0864-1_1

1.1 Non-linear Chemical Dynamics

Various types of oscillating behaviors such as emergence of chemical waves, chaotic patterns, and a rich variety of spatiotemporal structures are investigated in oscillatory chemical reactions in association with nonlinear chemical dynamics [1–3]. In non-equilibrium condition, the characteristic dynamics of such chemically reacting systems are capable to self-organize into diverse kinds of assembly patterns. With the help of nonlinear chemical dynamics, the complexity and orderliness of those chemical processes can be explained properly. Various biological processes which exhibited very time-based fluctuations especially when they are away from equilibrium have also been described by mechanistic considerations and theoretical techniques of nonlinear chemical dynamics [4–7].

Similar to many new areas of sciences, the nonlinear chemical system is emerged as a highly interdisciplinary branch of science and is characterized possibly by working functionality of well-known theories and their experimental data. Many examples of such systems are applied in almost all branches of chemistry including biology, where researchers have focused their two specific aims: (i) How system's behavior (concentration, energy etc.) vary with function of time or spatial coordinates? and (ii) What mechanisms are involved during the emergence of those spontaneous, coherent, often periodic, structural patterns? Recent efforts have been made primarily for the morphological stability of growing bodies [4], crystal growths [8, 9], non-equilibrium crystallization phenomena [10], self-oscillating gels [11], preparation of new materials [12, 13], surfaces, and autocatalytic reactions [7, 14–16].

1.2 Oscillatory Chemical Reaction

Oscillatory chemical reactions always undergo a complex process and accompany a number of reacting molecules, which are indicated as reactants, products, or intermediates. An elementary reaction is occurred by the decrease in the concentration of reactants and increase in the concentration of products. Initial concentration of the intermediates of such reaction is considered low, which approaches almost pseudo-equilibrium state in middle at this moment speed of production is essentially equal to their rate of consumption. In contrast to this, an oscillatory reaction undergoes with the decrease in the concentrations of reactants and increase in the concentration of the products. But the concentrations of intermediates or catalysts species execute oscillations in far from equilibrium conditions [1]. An oscillatory chemical reaction is accompanied by some essential phenomenology called induction period, excitability, multistability, hysteresis, etc. [1, 4]. These characteristic phenomena could be useful to determine the mechanism and behavior of the oscillating reaction.

$$A + B \quad \rightleftharpoons \quad X + Y, \ldots \quad \rightleftharpoons \quad P + Q$$

Reactant disappear   Intermediate either   Product appear
approaches to a pseudo−steady state
or execute oscillations

$$(1.1)$$

The oscillatory phases of those chemical reactions are driven by gradual decline in the Gibb's free energy of reacting media [1, 2]. The decreasing free energy could be found in the entire family of the chemical reactions; however, a limited numbers of reactions exhibit oscillating phenomenon. The oscillatory reaction should have some unusual features which allow showing some intermittent patterning behaviors. These characteristics can observe only when reactants react together and intermediates approach toward equilibrium position. The reaction mechanism for such reaction process is very complex because concentrations of some intermediates or catalysts change their states in periodic fashion which facilitates reaction into oscillatory phases.

Studies on the chemical oscillators are suggested that reaction mechanisms of such systems should have few common properties: (1) At the time when oscillation is occurring, the reacting media should be far-equilibrium condition. (2) The driving forces required for reaction system should operate in two different routes which intermittently switch from one route to other. (3) Out of two routes, one is producing certain intermediates which are totally consumed by ensuing reaction route. Function of intermediates triggers the reaction processes and simultaneously controls the reaction pathways by switching one to another. In other sense, at acute low concentration of the intermediates, the reaction proceeds through production route whereas it follows the consumption reaction route in the state of high concentration of intermediates. For required oscillation, the intermediate concentration is decreasing or increasing in rhythmic fashion for specific time duration.

1.3 Physical Criteria of Chemical Oscillations

Usually, the oscillation of chemical reactions has been analyzed through the thermodynamic and kinetic considerations. Thermodynamics of those chemically reacting system must be kept away from equilibrium and the free energy (ΔG) of those reaction is extremely negative. Besides this, the kinetic requirement is associated with the mass-action dynamics [17]. In the mass-action dynamics, the concentration of products and intermediates necessarily rises to power of (≥ 1), which attributes nonlinear dynamics. However, the substantial occurrences of nonlinearity in those dynamic laws do not alone guarantee chemical oscillations.

Generally, the nonlinearity is further relates with autocatalysis, self-inhibition, or a delayed feedback loop etc., which cooperatively acted to play role in chemical oscillations [1, 3].

1.4 General Non-equilibrium Thermodynamics

Thermodynamics which deals with systems that is essentially away from the equilibrium state is known as non-equilibrium thermodynamics. Many physical and chemical systems in the practical life are not in equilibrium state because they are continuously/discontinuously performing some actions where changes in their mass and energy from one form to another are taken place. Studies of such thermodynamical systems are required some added functionality other than basics of thermodynamic equilibrium systems. Several physical and chemical systems are remained beyond the reach of macroscopic methods of thermodynamics even nowadays. A major complexity for macroscopic thermodynamics is the definition of entropy which not comes exactly in the thermodynamic equilibrium [3, 5].

The idea of non-equilibrium and nonlinear dynamic laws can be understood by a simple method. Just assume a glass bottle filled with soft drink kept open on a bench. This state is called mechanical equilibrium condition. Slanting the bottle marginally produces a non-equilibrium state, which moves towards a new equilibrium state when drink flows out from the bottle. The Pepsi drink will run out gently in the condition of small angled. It may call a very close to equilibrium state. Energy required for flowing of Pepsi drink is small in above-mentioned condition. This can be seen in many physical and chemical systems in which all motions should have non-oscillating phase near to the equilibrium. Slanting the Pepsi drink into the further levels increases the distance from the equilibrium point. At this moment, evidently the Pepsi drink does not flow out gently. It splashes out and oscillation is taken place. Suitability and slanting at which oscillation is possible depend upon the geometry of the glass bottle. This can be measured through the dynamic laws and governing equations of the non-equilibrium thermodynamics. Similarly, chemical systems are also showing oscillation in non-equilibrium condition, which usually depend upon the governing equations of thermodynamics and dynamical laws.

1.5 Non-equilibrium Thermodynamics of Oscillatory Chemical Reaction

Chemical oscillations were not recognized at early periods of its discovery. Research community did not accept that chemical oscillation is possible and a genuine homogeneous oscillating reaction is impractical topic. Many scientists declared that studied reaction is not a consistent oscillatory reaction. They were argued that such phenomenon is due to heterogeneous artifacts of effervesce of bubbles or formation contaminated colloids [1–4]. The majority of disagreements were stated that such spontaneous self-organization phenomena are violating the 2nd law of thermodynamics [2].

The second law of thermodynamics is a revolutionary physical law which applied properly in all branches of sciences. It is exactly well-matched on the direction where the reaction changes spontaneously. This states that the entropy of the universe is continuously increases for a spontaneous change, i.e.,

$$\Delta S_{total} > O \tag{1.2}$$

For an open (or isolated) system (exchange neither energy nor matter with their surrounding), the entropy or ΔG changes monotonically in a spontaneous method. A scheme for two basic thermodynamic systems is presented in Fig. 1.1

An oscillating chemical reaction can be explained as a chemical model of working mechanical pendulum which passes through its equilibrium position. A chemical system never reaches to its equilibrium state when oscillation is occurred. Such reaction system can accompany a number of reacting components. It is assumed that at least two components are oscillating during each cycle of oscillation. The concentration of any one of those components should be near to its equilibrium position, while the remaining components must be far away from equilibrium point. The entropy increases by marginally in each cycle oscillation and the Gibbs free energy decreases monotonically. These two state functions bring the system far away from equilibrium which is monitored by the laws of non-equilibrium thermodynamics [18, 19].

First of all, the Prigogine explained that the chemical oscillations were possible in some chemical systems, provided they were far away from equilibrium [2]. The systems might be organized by decrease in their entropy which last until the gross entropy of the universe has become positive $\Delta S_{total} > O$. The concentrations of the intermediates in such a reaction could oscillate with the function of time, while the free energy gradually decreases. It can also be explained as the periodic fluctuation in concentration levels of intermediates is recompensed by the entropies of reaction processes involved in the oscillations as illustrated properly in Fig. 1.2. It has also been suggested that, an oscillator of chemical system is reasonably different than mechanical or electrical oscillators although a chemical oscillation is occurred in

Fig. 1.1 Two common types of thermodynamical systems

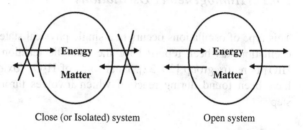

Close (or Isolated) system   Open system

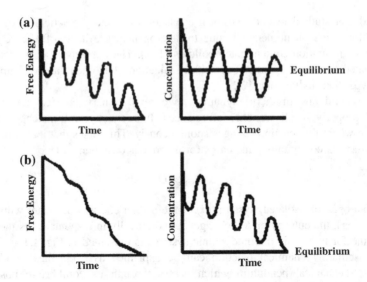

Fig. 1.2 Two types of plausible oscillations in closed systems. **a** Oscillations around equilibrium are conflicting of the second law of thermodynamics and **b** Oscillations on the way of equilibrium by decreasing in the free energy monotonically and consistent with the second law of thermodynamics (Adopted from Ref. [1])

far-equilibrium condition which is entirely monitored by laws of non-equilibrium thermodynamics. In a typical chemical oscillator, along with concentration fluctuation of intermediates the conversion of higher free energy reactant into a lower energy products has also been taken place is followed by monotonic decrease in free energy of systems [1].

1.6   Types of Chemical Oscillators

The following types of oscillations can be possible in chemically reacting systems which are briefly described below:

1.6.1   Homogeneous Oscillations

This type of oscillations occurs in a single physical state. The earliest homogenous oscillation has been reported for the catalytic reaction of hydrogen peroxide by HIO_3/I_2 redox group [20–22]. Dual roles of H_2O_2 (oxidizing and reducing agent) have been found during reaction which involves through the following reaction steps:

$$5 H_2O_2 + I_2 \rightarrow 2 HIO_3 + 4 H_2O \qquad (1.3)$$

and reduces HIO_3 to I_2:

$$5 H_2O_2 + 2 HIO_3 \rightarrow 5 O_2 + I_2 + 6 H_2O \qquad (1.4)$$

Thus, H_2O_2 apparently disappears as follows:

$$H_2O_2 \rightarrow H_2O + 1/2 O_2 \qquad (1.5)$$

1.6.2 Heterogeneous Oscillations

Some chemical reactions are insufficient for oscillations. But, they were coupled with reaction processes like layering structures in association with diffusion for development of chemical oscillations. Few reported examples of such reactions are enlisted below.

The rhythmic luminous patterns in vapor-phase oxidation of phosphorus have been investigated. This attributed an episodic consumption and regeneration of a protecting oxide layer [23]. Rayleigh [24] has found moving pulsation of luminosity in a tubular reactor between phosphorus and oxygen in two opposite inlets.

1.6.3 Thermochemical Oscillations

A reaction system which is not occurred in an isothermal condition and able to show chemical oscillations is driven by temperature changes that have significance for chemists in order to process controls and designing a reaction. Periodicity in such chemical oscillators which is certainly evolved by thermal changes has been investigated by several researchers especially in hydrocarbon fuels and their combustion derivatives [25]. A detailed theoretical and practical validation of those oscillating combustion reactions has also been reported in chlorination of CH_3Cl in vapor-phase reaction [26].

1.6.4 Electrochemical Oscillations

The most basic illustrations of chemical oscillation were investigated in electrochemical reactions. In the year of 1828, Fechner observed chemical oscillation in reactive electrodes. Nowadays several thousands of studies of chemical oscillation by means of potentiometric and galvanometric conditions have already published [27, 28].

1.7   Representative Oscillators in Chemical Systems

The well-characterized chemical oscillators should have following elements which are effectively control the phase and rhythm of oscillations:

Mechanistic section: It is a set of elementary reaction steps which is essential to describe how reactants form intermediates and also how these intermediates further combine with others to form ultimately some stable products.

Equation rates: A typical chemical oscillator consists of numbers of transitory reactions which can be explained with their differential equations. It is suggesting that the rate of change is taken place for all associated reactants, intermediates, and products during the oscillations.

Integral equation rates: This section explains how the concentration of reactants, intermediates, and products of oscillating reaction are varied in a function of time. These are acquired through the integration of differential equation rates.

Theoretical discussion of oscillatory reaction began since the discovery of the most famous Lotka's model in 1921. After the non-equilibrium theories of Prigogine (1968), many theories to explain the oscillatory chemical reactions have been proposed. A brief summary of some important oscillator models have been described below.

1.7.1   Oscillating Model of Lotka–Volterra

It is the most basic and properly explained model of oscillating reactions. In 1921, Lotka proposed a model to explain some oscillatory phenomena in biological systems [29]. This model composed numbers of sequential steps which are presented in Table 1.1. As suggested, the each step referred to a specific mechanism in which the reactant molecules come together to form some useful intermediates and products.

Table 1.1 Lotka–Volterra model and rate equations

Reaction step	Molecular reaction	Differential rate equations
1	$A + X \rightarrow 2X$	$\dfrac{d[A]}{dt} = -k_1[A][X]$
		$\dfrac{d[X]}{dt} = -k_1[A][X]$
2	$X + Y \rightarrow 2Y$	$\dfrac{d[X]}{dt} = -k_2[X][Y]$
		$\dfrac{d[Y]}{dt} = -k_2[X][Y]$
3	$Y \rightarrow B$	$\dfrac{d[Y]}{dt} = -k_3[Y]$
		$\dfrac{d[B]}{dt} = k_3[Y]$

In order to represent oscillating model in equation, it can be suggested that a reactant molecule of A reacts with a molecule of X forming two molecules of X. Above-mentioned reaction consumes a unit of molecule A by adding a molecule X. Rate of reaction is directly proportional to the concentrations of molecules A and X.

Where k_1, k_2, and k_3 are the reaction rates. A and B are the reactant and the product, whereas X and Y are the two transient intermediates, respectively. The gross rates of reactant A, product B, and the intermediates X and Y are obtained by calculating from the inputs of each contributing step.

$$\frac{d[A]}{dt} = -k_1[A][X] \tag{1.6}$$

$$\frac{d[X]}{dt} = k_1[A][X] - k_2[X][Y] \tag{1.7}$$

$$\frac{d[Y]}{dt} = k_2[X][Y] - k_3[Y] \tag{1.8}$$

$$\frac{d[B]}{dt} = k_3[Y] \tag{1.9}$$

The reaction steps 1.7 and 1.8 are autocatalytic in nature, because the X and Y accelerate their own productions. These two reaction processes are also significant for the evolution of the oscillatory cycles. The model was very effective to explain some biological interactions between the ecosystems. This model is also known as predator–prey theory.

1.7.2 Brusselator Model of Oscillating Chemical Reactions

This oscillating system was investigated in 1968 by Prigogine and Lefever. The partial modification in the form of "Brusselator" was done by Tyson group at Free University of Brussels in 1973 [30]. The mechanism of this oscillating system along with the rate of equations is properly shown in Table 1.2.

Where k_1, k_2, k_3, k_4 are the rate constants of corresponding reaction steps. The main reaction is occurring in the form of, $A + B \rightarrow C + D$ along with the transitory emergence of the intermediates X and Y.

1.7.3 Oregonator Model of Oscillating Chemical Reactions

This oscillating system has been investigated by Field and Noyes in 1974 [31]. A simplified form of Oregonator model is presented in tabulated form in Table 1.3.

Table 1.2 Brusselator model and rate equations

Reaction steps	Molecular reaction	Differential rate equations
1	$A \rightarrow X$	$\dfrac{d[A]}{dt} = -k_1[A]$ $\dfrac{d[X]}{dt} = k_1[A]$
2	$2X + Y \rightarrow 3X$	$\dfrac{d[X]}{dt} = k_2[X]^2[Y]$ $\dfrac{d[Y]}{dt} = -k_2[X]^2[Y]$
3	$B + X \rightarrow Y + D$	$\dfrac{d[B]}{dt} = -k_3[B][X]$ $\dfrac{d[X]}{dt} = -k_3[B][X]$ $\dfrac{d[Y]}{dt} = k_3[B][X]$ $\dfrac{d[D]}{dt} = k_3[B][X]$
4	$X \rightarrow E$	$\dfrac{d[X]}{dt} = -k_4[X]$ $\dfrac{d[E]}{dt} = k_4[X]$

Where k_1, k_2, k_3, k_4, and k_5 are the reaction rates. The effective reaction is resulted through combining of reactions 1, 2, 4, twice of steps 3, 5, respectively, and can be simply represented as $A + 2B \rightarrow P + Q$. By incorporation of autocatalytic reaction steps, the Oregonator model has become the most effective mechanism to explain even much complex oscillatory behaviors of reactions.

1.8 Belousov–Zhabotinsky (BZ) Reaction

1.8.1 Discovery of BZ Reaction

The discovery of BZ reaction was made early in 1950 by Belousov [32] which evolved a prototype system for investigating self-organization phenomenon in reaction–diffusion system. Actually, Belousov was more interested for the modeling of Krebs cycle containing inorganic system to reproduce the functionality of metal ions bonded protein molecules analogue to enzymatic systems. This research leads to an unexpected investigation, and later, it was accepted as earliest paradigm of true oscillating reaction [33, 34]. During the experiment, Belousov observed that citric acid, acidic bromate, and cerium ions oscillated periodically between colorless and yellow, which resulted the production of carbon dioxide. At the beginning, the incredulity of BZ reaction was congregate in the form of perception of an oscillating chemical reaction which has considered being in contravention of the second

Table 1.3 Oregonator model and rate equations

Reaction steps	Molecular reaction	Differential rate equations
1	$A + Y \rightarrow X$	$\dfrac{d[A]}{dt} = -k_1[A][Y]$ $\dfrac{d[Y]}{dt} = -k_1[A][Y]$ $\dfrac{d[X]}{dt} = k_1[A][Y]$
2	$X + Y \rightarrow P$	$\dfrac{d[X]}{dt} = -k_2[X][Y]$ $\dfrac{d[Y]}{dt} = -k_2[X][Y]$ $\dfrac{d[P]}{dt} = k_2[X][Y]$
3	$B + X \rightarrow 2X + Z$	$\dfrac{d[B]}{dt} = -k_3[B][X]$ $\dfrac{d[X]}{dt} = k_3[B][X]$ $\dfrac{d[Z]}{dt} = k_3[B][X]$
4	$2X \rightarrow Q$	$\dfrac{d[X]}{dt} = -k_4[X]^2$ $\dfrac{d[Q]}{dt} = k_4[X]^2$
5	$Z \rightarrow Y$	$\dfrac{d[Z]}{dt} = -k_5[Z]$ $\dfrac{d[Y]}{dt} = k_5[Z]$

law of thermodynamics. The main reason was mutual transformation of products to reactants into intermediates to oscillate at its equilibrium state. This was also associated with Gibbs free energy of the system, which is totally contrary to the second law arises [35]. Although the oscillatory phenomenon arises from a chemical reactions which is not an equilibrium process, oscillations in the concentration of intermediates are not excluded by thermodynamics, as the system does not oscillate about its equilibrium free energy point, rather it oscillates far from equilibrium. As a result, the net free energy is always going to be decreased on the direction of its equilibrium point, gratifying the condition of the second law of thermodynamics. After some time, Anatol Zhabotinsky reconstituted the original recipe of the Belousov by substituting citric acid with malonic acid (MA), and changing the cerium catalyst with ferroin which is a redox indicator with better color appearance. The ferroin-catalyzed reaction has come to symbolize the modified BZ reaction in its most significant form [36]. Toward after that in nonlinear chemical dynamics, the ferroin-catalyzed BZ reaction has widely studied to exhibit a number of phenomena exemplary of an oscillating chemical reaction for the study of nonlinear chemical kinetics.

1.8.2   Oscillatory Behavior in the BZ Reaction

The temporal oscillating patterns of certain chemical intermediates have been observed only in a stirred BZ reaction system. Similar to cerium-catalyzed BZ reaction, the oscillation is occurred between colorless and yellow color at an assured time interval. There are some other important indicators where oscillations can be monitored due to the gradual color change, either directly in batch reactor [37, 38] or via spectrophotometric measurements [39, 40]. The most excellent illustration of oscillations manifested in batch reactors in the form of color variation of ferroin-catalyzed BZ reaction system. On the other hand, in some other reaction systems such as the manganese-catalyzed system and the cerium-catalyzed reaction, oscillations can be observed with the help of UV–visible spectrophotometer [41] where change in color might be monitored less distinctly.

The important techniques for measuring the oscillations in the chemical reaction are the potentiometry. The advantage of this technique is that by using a bromide ion-sensitive electrode, the composition of bromide ions in the reaction system can be monitored easily. However, platinum electrode is susceptible to changes in the oxidation state of the metal-ion catalyst. The measured electrode potential can be used to monitor oscillations in $[Br^-]$ and $[M_{ox}]/[M_{red}]$ with the help of a suitable reference electrode.

1.8.3   Mechanism of BZ Reaction

The modified BZ reaction involves the oxidation of MA by acidic bromate occurred in the presence of redox metal catalyst having a one-electron ion couple [42]. The FKN [43, 44] is the key concept to understand the mechanism of the BZ reactions. From the chemical kinetics point of view, general investigation of the reaction system has been revised and modified which was reported in an article published in 1972. But many advantages and specific properties of the BZ reaction has been added. In comparison with the beginning of the BZ reaction, now its mechanism can be easily understood with the help of FKN.

1.8.3.1   The Chemistry of BZ Reaction

Field, Koros, and Noyes (FKN) [45] developed the mechanism of BZ reaction in 1972. Basically FKN model containing a total of 11 reaction chains involving 15 different intermediates and molecules [42]. The mentioned reaction chains were additionally formulated into three major processes [46]. The first processes starts with the depletion of bromide ions by a reaction of bromate ions. During this processes Br_2 is produced and reacts with MA to form bromomalonic acid. In the oscillatory phenomenon of the reaction this key intermediates then reacts with the oxidized form of the metal-ion catalyst to "reset the clock." The second processes generate the production of bromous acid ($HBrO_2$) via the autocatalytic processes and the oxidation of the metal-ion catalyst.

The Belousov–Zhabotinsky (BZ) system is a methodically characterized chemical oscillation and provides an archetype scheme for study of wide ranges of patterning features in oscillatory chemical reactions [47–53]. This consists of bromination reaction initially and auto-oxidation of organic substrates is takes place in sequential processes by bromate ions. Overall, the reaction is catalyzed by redox catalysts in a concentrated water-acidic solution.

The BZ reaction showed two fascinating behaviors: (1) It gives temporal oscillations in wide range of reaction conditions. (2) They also exhibited various orders of periodic and spatial pattern formation, both in time scales and space coordinates. This intriguing phenomenology has been studied by a numbers of techniques and experimental conditions [1, 3, 4].

The BZ oscillating reaction has been discovered by B.P. Belousov, in 1951. In his early attempt, he wanted to make a chemical model for Krebs cycles (an energy pathway). For experimentation, he used $KBrO_3$, acidified solution of citric acid and a metal catalyst (Cerium (IV) metal ions). A redox indicator has also been used for observing the end of the reaction-phase. They found that the citric acid is oxidized into CO_2, and BrO_3^- is reduced to Br^- ion. He observed a periodic color changes for duration of 10 min. He wanted to publish this report but editor refused his proposal and explanation because, it seems to be contrary with 2nd laws of thermodynamics.

In 1961, Anatol Zhabotinsky was looking for Belousov's reaction. He replaced citric acid by malonic acid and found similarity in reaction behavior [52]. Zhabotinsky has proposed a reaction equation (Eq. 1.9) as shown below,

$$3\,CH_2(CO_2H)_2 + 4\,BrO_3^- \rightarrow 4\,Br^- + 9\,CO_2 + 6\,H_2O \qquad (1.10)$$

During the period of 1970 the Belousov–Zhabotinsky reaction was not well understood but it remains an exciting topic and laboratory curiosity for researchers.

In 1972, Field, Koros, and Noyes (FKN group investigated the kinetic details which characterized the entire aspects of BZ reaction, illuminating the essential constituents of reaction and their roles in oscillations [31, 53]. The kinetic model composed three consecutive reaction processes as, (A), (B), and (C). The process A is a fast reaction step, the process (B) is an autocatalytic set of reaction, and C is the process where (Br^-) ions are consumed. The oxidation of metal catalyst ions has also been taken place in the processes (A) and (B), respectively. A recovery step (process C) involves for the reduction of metal catalysts and regenerates the necessary reactant (Br^-) ions for re-initiating the oscillatory-phase reaction from beginning. A schematic model for description of the chemistry of BZ reaction is shown in Fig. 1.3.

Fig. 1.3 Three reaction
processes involved in the BZ
reaction

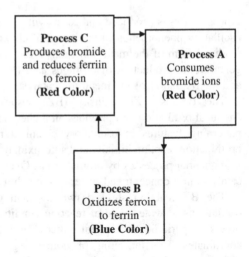

Process A: Removal of Br^-

$$BrO_3^- + Br^- + 2H^+ \rightarrow HBrO_2 + HOBr \tag{1.11}$$

$$HBrO_2 + Br^- + H^+ \rightarrow 2HOBr \tag{1.12}$$

$$HOBr + Br^- + H^+ \rightleftharpoons Br_2 + H_2O \tag{1.13}$$

Process B: Self-catalytic reaction of $HBrO_2$ and oxidation of Ce^{+3}/Fe^{+2} catalysts which is measured by color changes or potentially generally in two forms, ferroin (red) and ferriin (blue)

$$BrO_3^- + HBrO_2 + 2H^+ \rightarrow 2BrO_2^- + H_2O \tag{1.14}$$

$$BrO_2 + M^n + H^+ \rightarrow HBrO_2 + M^{n+1} \tag{1.15}$$

$$2HBrO_2 \rightarrow BrO_3^- + HOBr + H^+ \tag{1.16}$$

Then,

Process C: Rearranging the periodicity by the reduced state of Ce^{+3} or Fe^{+2} and emergence of bromide ion

$$2M^{n+1} + 2BrMA \rightarrow Br^- + 2M^n + Other\ products \tag{1.17}$$

Bromination of MA has been taken place spontaneously and BrMA is formed:

$$MA + Br_2 \rightarrow BrMA + Br^- + H^+ \tag{1.18}$$

1.8.4 Methodology and Analytical Techniques

Under thermodynamic and kinetic conditions, the oscillatory reactions have been studied in many different ways [54–58], which could show following periodic behaviors:

(1) Visually from sharp color changes,
(2) Potentiometrically from change in potential,
(3) Amperometrically from change in current,
(4) Spectrophotometrically from change in absorption,
(5) Thermometrically from change in temperature,
(6) pH-metrically from change in pH,
(7) Proton resonance (NMR), magnetic resonance (MRI), and electron resonance (EPR) methods have also been used to study the oscillatory reactions.

The chemical oscillations start with a short induction period that undergoes for few second to minutes. Depending upon the conditions and the nature of the reactions, specific method has been used to measure the oscillating behaviors. Typical example of a reaction system with different oscillating property has been presented in Fig. 1.4.

1.8.5 Conditions and Factors Affecting the Belousov–Zhabotinsky Reaction

The oscillating reaction exhibits a complex dynamic behavior and wealthy pattern characteristics during their reaction process. The reaction process is very sensitive to used chemical reagents, temperature, thickness of solution layer, mixing, or stirring property etc. The reaction condition and some factors that control the oscillation and pattern formation are described below:

1.8.5.1 Chemical Components

Organic Substrate: The substrate molecules react slowly with acidic bromate, and they should have brominable and oxidizable chemical groups.
Catalysts: They have at least two oxidation states which differ by an electron. They should have standard reduction potential between 1.0 and 1.5 V (e.g., Ce^{4+}/Ce^{3+} Mn^{3+}/Mn^{2+}; $Ru(bpy)_3^{3+}/Ru(bpy)_3^{2+}$; and $Fe(phen)_3^{3+}/Fe(phen)_3^{2+}$).
Oxidizing Agents: Bromate is excellent candidate for oxidizing agent in entire family of chemical oscillator, since they participate in the reaction by two means: (i) two-electrons (oxygen transfer) reaction steps and (ii) one-electron reaction steps.

(a)

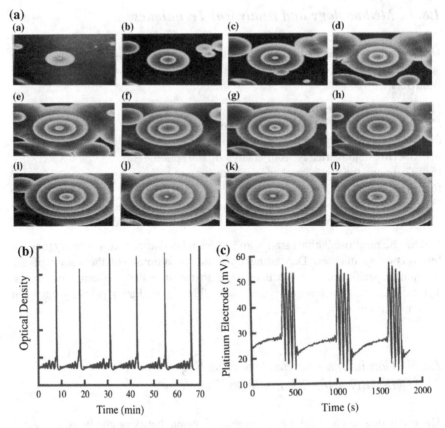

Fig. 1.4 **a** Emergence of chemical waves in unstirred medium [48], **b** and **c** The temporal oscillations with respect to optical density, and change in potential in well-stirred BZ reaction medium (Adopted from Ref. [1])

1.8.5.2 Thermodynamic and Kinetic Consideration

The BZ reaction occurs in irreversible manner and thermodynamically characterizes as extremely low negative Gibb's free energy (ΔG). This is equivalent to entropy changes between the systems and surrounding. The entropy-produced or gradual energy dissipation process facilitates reaction to form space–time structures or dissipative patterns. The reaction never reaches to equilibrium state as long as dissipation of energy occurred. The reacting system when comes, at or close to equilibrium position, they will stop oscillations automatically.

The kinetics of oscillatory chemical reaction must have nonlinearity and their rate equations be supposed to quadrant function of concentration of reactants. The kinetically control reaction steps have two values, (+ve) is associated with auto-catalysis and (−ve) value is associated with auto-inhibition feedback loops [1].

1.9   Chemical Wave and Pattern

1.9.1   Pattern in Nature

In the nature, pattern formation phenomenon is a frequent and exciting subject matter throughout. Patterns formation through biological, chemical, and physical processes is a profound example of dissipative structure in a diverse range of phenomena [59–61]. The common example of pattern formation in nature at small-scale systems includes the growth of colonies of bacteria; pattern in the coat of animals such as strips and dots in zebra, tiger, and fish; and the electrical impulses that control the heart beating. However, at larger-scale systems, pattern formation includes cloud formations, networks of sand dunes, plankton blooms observed in the ocean, and collections of stars into galaxies. The main motivation for the study of pattern formation in chemical reactions is that to apply this key concept to an understanding of how chemical systems may be used as model pattern forming phenomenon observed in nature. One can simply compare the formation of spiral wave pattern as could be seen in slime mold *Dictyostelium discoideum* and spiral wave pattern formation experimentally in the classic BZ reaction (Fig. 1.5c, d). It would be appreciated how these two pattern formation in two different pathways might be linked together. However, the pattern formation in a similar way at two different species may or may not be given the required underlying mechanism to understand these interesting facts.

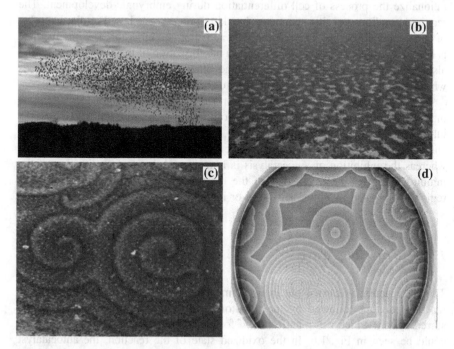

Fig. 1.5 Self-organization phenomenon at different spatial scale **a** flocking, **b** patchy vegetation in Nigeria **c** spiral pattern in *Dictyostelium discoideum* and **d** BZ reaction performed in Petri dish [60]

1.9.2   Traveling Wave in RD System

In an excitable chemical reaction system, the pattern formation generates due to the coupling of reaction kinetics with diffusion of reactant species, known as a reaction–diffusion system. As a result of the localized production of the autocatalyst, if a discrete point in the excitable medium is occur, then the autocatalyst would diffuse into neighboring regions inside the reaction mixture, resulting it becomes excited. Due to happening of this phenomenon in the reaction system a traveling front of excitation propagating outwards from the initial point of perturbation. Therefore, the autocatalyst is usually called "propagator" species [60]. In some special cases, the system may come back to its initial state by another chemical species therefore called "controller" species. In that condition, a wave is generated, provided the system has improved from its intractable state.

1.9.3   Stationary Turing Structures

Alan Turing, a mathematician, in his research paper entitled *The Chemical Basis of Morphogenesis* [62], had mentioned that the coupling of autocatalysis and diffusion could generate stationary patterns at the starting phase, and however, this system was hypothetical. Turing realized that this kind of reaction could be used to rationalize the process of cell differentiation during embryonic development. The required condition for this symmetry–breaking process was based on the interplay of two chemical species: These species are activator and inhibitor. The intermediate is known as activator species which assists the formation of the second species and also catalyzes its own production. On the other hand, the intermediate species which inhibits the formation of the activator is known as inhibitor.

The fundamental required condition for the formation of stationary patterns is that the inhibitor species should be able to diffuse much more quickly. Thus, the inhibitor species would felt at a longer range than the activator species. The autocatalytic reaction and inhibitor production species are localized phenomenon, whereas the inhibition of autocatalysis is long-ranged due to fast diffusion of the inhibitor causing spatial ordering of the reaction medium with various different regions dominated by either activation or inhibition of the reaction.

1.9.4   Traveling Wave

The BZ reaction is carried out in two-dimensional, i.e., Petri dish filled with the reaction mixture followed by the monitoring of oscillatory phenomenon in the stirred system which can manifest itself formation of traveling chemical waves as could be seen in Fig. 1.6. In the oxidized state of the reaction, the autocatalyst

Fig. 1.6 The formation of traveling waves when the concentration of lipid is increased in a BZ reaction [66]

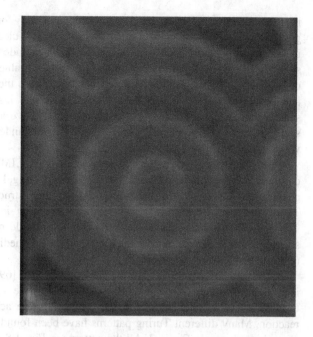

species $HBrO_2$ causes a wave front due to the diffusion which propagates through the reaction medium. Then, the oxidized metal-ion catalyst reacts with MA to restore the system back to its initial state at the back side of the wave front. Zhabotinsky [36] observed this interesting phenomenon in a 2-mm layer of the ferroin-catalyzed BZ reaction. At the initial phase, the pattern slowly organizes in the form of wave front. The coupling of multiple waves generates in the whole Petri dish forms relatively complex structures either in the presence of air bubbles or in the presence of dust particles called "target waves." Figure 1.6 represents the formation of rotating spiral waves [63–65].

1.9.5   Turing Pattern

When a reaction is carried out under the condition where the thickness of the solution mixture is less, the colored spatial patterns arise due to the variations in chemical concentrations and this pattern depends on the time. The other important pattern which is stationary in time and periodic in space, or periodic in both time and space is known as Turing structures. Turing pattern looks like very beautiful and has close resemblance with biological and chemical systems.

Diffusive instabilities usually require a special relationship between diffusion coefficients. Ordinarily, for Turing instability the diffusion coefficient of the inhibitor, D_{inh}, should be significantly larger than that of the activator, D_{act}. This theoretical relation applies, however, only in the absence of cross-diffusion [66], i.e., the diffusion of each species must depend only on its own concentration gradient and not on gradients of other species, and it is clearly an oversimplification in systems with more than two variables, which include essentially all biological systems. In the most real systems, there are many variables, and it is often difficult or impossible to find a model that adequately explains the phenomenon of interest and possesses only two variables. Nevertheless, the insight from theory that a system that exhibits Turing-type behavior should have at least one fast-diffusing species that serves as an inhibitor, i.e., prevents or hinders autocatalysis, remains valid.

Turing structures have been widely studied in CIMA reaction [67] and its derivatives. When the CIMA reaction is performed in gel media by using starch as an indicator, some striped and hexagonal (spotted) structures [68] are observed. These structures are shown in Fig. 1.7. First, a starch–iodide complex is formed. The activator (iodine species) and the starch–triiodide complex generate Turing structures which diffuse much more slowly in the gel medium than inhibitor species (chlorite or chlorine dioxide).

In another important study in the BZ-AOT system [69], the species Br_2 can be identified. Bromine diffuses rapidly in the oil phase and inhibits autocatalysis through its facile conversion into Br^-, which is the actual inhibitor in the BZ reaction. Many different Turing patterns have been found in the BZ–AOT system; several are shown in Fig. 1.8. All the patterns in Fig. 1.8 are stationary, even those in Fig. 1.8f that resemble the concentric ring wave patterns. In Fig. 1.8, we can see spots (b), stripes (a), and labyrinthine (d) patterns. Mixtures of stripes and spots are also possible as shown in Fig. 1.8e.

Fig. 1.7 Formation of Turing structures: hexagonal (*left*) and stripe (*right*) [66]

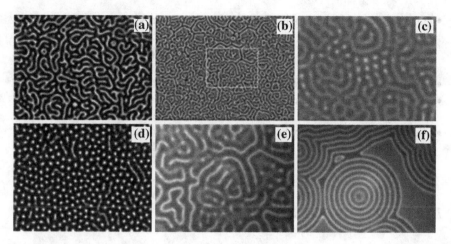

Fig. 1.8 Turing patterns in the BZ–AOT system [69]

1.10 Microemulsions

A thermodynamically stable mixture which consists of water, oil, and a surfactant molecule is known as a microemulsions system. Microemulsion system is the example of formation of Turing pattern in the reaction–diffusion system currently. The main advantage of the microemulsion system is that the surfactant molecule provides a stabilizing interface between two other immiscible phases [70]. Due to the thermodynamic stability, it distinguishes them from a normal emulsion [71]; emulsion mixtures consist of oil and water; however, such mixtures will gradually separate into two phases and remain unstable. On the other hand, it represents a configuration of the water, oil, and surfactant which corresponds to energetically favorable. Transparency is another distinguishing feature of the microemulsions system.

The AOT microemulsion represents one of the most exciting chemical reaction–diffusion systems for the formation of Turing structure, reported by Epstein research group [72]. Some nonpolar solvents such as octane, decane, and cyclohexane also form AOT microemulsions. The advantage of the AOT-micromulsion system without the need for a cosurfactant [72] is its ability to solubilize large amounts of water. The structure of AOT surfactant molecule consists of a polar, negatively charged head group, and two long, hydrophobic tails.

1.10.1 Applications of Microemulsion

Various types of microemulsion systems can form with neutral, anionic, cationic, and even zwitterionic surfactant molecules, with unique properties and characteristics. Some important applications of microemulsion are as follows:

- One of the significant applications of microemulsion systems is in the field of EOR. It can be used for implantation of various techniques for increasing the amount of crude oil that can be extracted from the ground water of oil source [73].
- During the machining processes, it can also be used as lubricants [74].
- One of the most important applications of microemulsion systems is that of drug delivery.
- Some other important applications include reduced toxicity, enhanced uptake, and the ability to control the rate of release of the drug into the blood [73].

1.10.2   The BZ–AOT System

Balasubramanian et al. in 1988 [75] reported first illustration of the inclusion of the BZ reaction into an AOT reverse micelle system. The coupling of an oscillating chemical reaction which shows spatial and temporal phenomenon relevant to bio-logical systems was the main motivation for this study. In manganese-catalyzed reaction system oscillatory behavior was monitored for this particular case. Vanag et al. [76] has been studied the BZ–AOT reaction in a great detail emphasized in particular on the formation of non-equilibrium chemical patterns.

1.11   Universality of Self-organization in Natural Science

Many phenomena in living systems involve periodic changes. The growth of plants or the development of embryos needs complex reaction processes where constituent atoms or molecules may form highly organized objects of scientific interests. The self-repeating dynamical events have often been observed in oscillatory reaction media with respect to time or space. The coupling between reaction–diffusion mechanisms in those dynamical systems can create self–spatial organization [77–80]. The natural systems exhibit two main characteristics during self-organization, as described below:

(1) They are thermodynamically open systems which can display a regular fluc-tuations and continuously exchanges energy and matter during their reaction process. The example includes entire natural and numerous synthetic objects.
(2) The resultant organizations which are associated with complicated nonlinear network of molecular interactions.

Fig. 1.9 Schematic graphics that showing the dynamical relationship between microscopic and molecular interactions

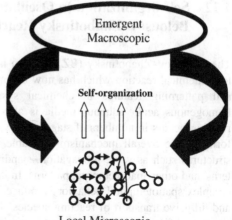

Consequence of self-organization process, the dynamical system enables to create intricate structural patterns of various shape and geometry. A long-range macroscopic interaction in space and time has been considered in those dynamical patterns.

An effort was made to visualize the self-organization phenomenon in a sketch which is shown in Fig. 1.9. It is described the molecular dynamic interactions between the early pattern forming state at microscopic scale and growth of complex structures at macroscopic scale. Vertical arrows specify that how the molecular dynamics acted to organize systems spontaneously into typical macrostructures. Concurrently, the bottom lines indicate that novel macroscopic systems are resulted from microscopic interfaces into additional levels of hierarchy. The driving forces acted for both microscopic and macroscopic processes are facilitating the emergence self-organizing patterns of required stability.

Some significant phenomena of self-organization in chemical, physical, and biological systems are as follows:

Chemical processes: Oscillating reactions, chain reaction of catalysis, molecular self-assembly, ordering in liquid crystals, monolayers self-assembly, molecular-phase transition, Langmuir–Blodgett films, etc.

Physical processes: Order–disorder structures, ordered-phase transitions, symmetry breaking, spontaneous magnetization, non-equilibrium crystallization phenomena, percolation, electrodeposition, formation of dissipative structures, turbulence and instabilities in fluid dynamics, and diffusion-limited aggregation process.

Biological processes: Excitation in muscles, pulsation of heart, calcium waves, natural fold-up of protein molecules, deposition of lipid bilayers, auto-regulation of homeostasis morphogenesis, hyper-cycles and autocatalytic networks, etc.

1.12   Self-organization in Oscillatory Belousov–Zhabotinsky Reaction

The Belousov–Zhabotinsky (BZ) reaction is a systematically exemplified oscillatory chemical reaction which has now become paradigmatic for nonlinear kinetics and patterning features in chemical systems. This reaction takes place in homogenous acidic-aqueous media is also catalyzed by redox metal ions. The reaction occurs in numbers of stages, and various transitory intermediates are also formed. The overall mechanisms are able to form a rich variety of space–time structures such as wave propagation, standing waves, target patterns, Turing patterns, and other spatiotemporal patterns. In the prototypical case of BZ reaction, the complex spatiotemporal behavior is induced by the interplay of nonlinear kinetics and diffusive transport of reacting species. The pattern formation process is called reaction–diffusion mechanism is introduced theoretically by Turing [5]. Morphogenesis of biological system and development of animal skin patterns can be also explained on the basis of reaction–diffusion mechanism of oscillatory reaction systems.

Pattern formation in oscillatory reacting media provides some representative and exciting examples of self-organization in chemical systems. Oscillating chemical system is regarded as dissipative systems or open systems, where space–time structures formed spontaneously at far from thermodynamic equilibrium [1, 2]. Self-organization occurs in dissipative systems or open systems which are often in far away from equilibrium by which it maintained the energy regulation. By exchange of matter and energy with environment, such systems are able to export entropy so that ordered patterns can emerge spontaneously.

The dissipative structures and spatiotemporal patterns formed in oscillating chemical systems commonly at far from equilibrium conditions; thus, they are also referred as non-equilibrium patterns. There are two different ways by which dissipative structures have been formed: (1) If any property (concentration, color, energy, entropy, etc.) oscillates uniformly with time which leads to temporal oscillations or temporal patterns, these patterns are very difficult to analyze directly. (2) If system's property, especially their physical states, is found to change in oscillatory (periodical) fashion with the functions of both time and spatial geometries, the natures of oscillation may come in the forms of moving fronts, pulsating waves, or more likely typical chaotic types of spatiotemporal structures. Once the system's property stops to oscillate, a sequence of stationary pattern emerged. These stationary patterns are described as Turing structures. For case 2, where system behavior changes aperiodic mannered, the resulted patterns in such condition are called chaotic patterns.

An intriguing example is BZ–AOT experiment [12], in which the constituents of the BZ system are dispersed in a reverse microemulsion containing an oil (octane) and the surfactant sodium bis(2-ethylhexyl) sulfo-succinate. Manipulation of

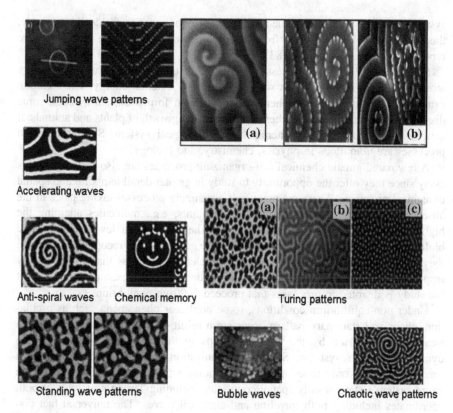

Fig. 1.10 Chemical waves and space–time structural patterns studied in the BZ-AOT reaction system (Adopted from Ref. [12])

chemical properties and remarkable array of patterns have been investigated through changing of compositions of surfactant/oil/water. Overviews of those pattering behaviors are shown properly in Fig. 1.10.

1.13 Self-organization in Material Science

Self-organization phenomena at nanoscale in materials science are mostly observed in soft matter component or supramolecular materials [81, 82] where some forces including electrostatic, short-range van der Waals, and dipolar usually stabilize the nanostructures. Sometimes, branched structures formed in metallic nanostructured systems but always depend on the use of diblock copolymer scaffolds. For example, the nanochains which are made from Au–Nps via self-organization processes were reported by Lopes and Jaeger [83].

Self-organization is the process of formation of ordered nanostructure by using atoms as a small building block without use of any external energy. The common example of self-organization phenomenon is cloud formation, snowflakes, some dissipative structure formed in chemical system and growth of plants and animals. It can be formed in physical, chemical, and biological system. Self-organization processes are ubiquitous in physics, chemistry, and biology.

A few paradigmatic chemical self-organizing processes are also included in this essay since they offer the opportunity to study in greater detail temporal and spatial dynamical processes. There are many self-organizing processes taking place in the absence of systematic non-equilibrium constraints, e.g., molecules attaining the highly ordered crystalline state. Included in the essay are a few phenomena in biology that do take place near equilibrium, e.g., molecular recognition, because they are indispensible for understanding the elementary steps that are ultimately responsible for the phenomenon observed. Aside from these examples, the bulk of the study is devoted to processes that proceed far from equilibrium.

Under non-equilibrium conditions, some nonlinear phenomena such as oscillation, chaos and stationary pattern occur are a result of the loss of stability by the steady states, caused by the feedback loops in the processes determining the dynamics of such systems. Such self-organization can be obvious itself as a function of either only time coordinate including simultaneous oscillations of the entire system's state or only spatial coordinate including Turing structures or both coordinates including both traveling and chemical waves. The universal fact discovered of such phenomenon in different systems is remarkable in the context of mathematical description.

In recent years, the basic understanding of such nonlinear phenomenon studied has been described. The changes in different steady states such as bistability and tristability during the electroreduction processes of nickel (II) derivatives complex compound of mercury ions at Hg electrode will act as the examples of the oscillatory changes. The universal self-organization scheme of the convective cells into hexagonal and fingerprints-like patterns will be shown for the electrohydrodynamic convection in the thin-layered, electrolytic cells, with the luminescence distribution showing the geometry of the fluid flows. Finally, the traveling chemical waves will be generated due to the catalytic oxidation of thiocyanates with hydrogen peroxide. The various applications of such kind of nonlinear phenomena under the scope of material science will be emphasized further based on the some interesting selected example. These examples will cover along with others, the effect of the self-organized convection on the morphology of the surface of cathodically deposited copper, the formation of other surface (micro) patterns, as well as polymerization reactions.

1.14 Applications of Self-organization in Materials Science

In the process of synthesis of materials with desired properties, all kinds of dynamical instabilities can manifest themselves. These instabilities arise some fundamental questions in mind: Whether these instabilities create nuisance or are they useful in the synthesis? Actually both can be possible. In a large industrial reactor, if the process starts to proceed with oscillations against our desire, it results in the loosing of our control over the reaction course, leading to the damage of the reactor in the extreme case. Although, if someone is familiar with nonlinear dynamics and its effect in negative aspects, then such undesirable situation can be avoided by optimizing the reactor construction or reaction conditions. At the same time, we can acquire evident advantages from dynamical instabilities during new material designing. There are plenty of examples of such situations in polymer chemistry [84, 85] and electrochemistry. In polymer chemistry, during polymerization reactions, we can distinguish the role of nonlinear dynamics between two essentially different aspects [86]. One polymerization process itself can exhibit nonlinear instabilities through intrinsic nonlinearities and feedback loops. In contrast, there are such polymerization reactions, in which instead of exhibiting instabilities, coupled with the oscillatory chemical system, e.g., the famous BZ process. In the following sections, these two cases will be explained briefly.

1.15 Oscillations and Self-organization in Material Science

Oscillatory chemical systems and intriguing patterns are highly fascinating and drawn a considerable interests for global scientists in last decade due to their scientific importance in material science. The studies on detail mechanisms of attractive space–time structures suggested that such patterning characteristics are common in diverse material processing, stability of growing body and non-equilibrium crystallization processes [87–89]. Under instability conditions or non-equilibrium states, many physical processes give rise variety of spontaneous dissipative structures, as suggested for oscillatory chemical systems. Basics of oscillating system that have been demonstrated in several fluidic media are attributing the pattern formation by the effects of motion of convection, periodic precipitation, periodic/aperiodic oscillations, chaotic patterning, ordering of oxidized SCN^-/H_2O_2 patterns, as well as the bistability and tristability in electrochemical reduction of azide complexes with nickel (II).

Proper description of nonlinear science in material science has been demonstrated in various crystallization processes of polymeric systems [90, 91]. As those systems were found to be capable to show oscillation independently by intermolecular interfaces or transitory oscillating phases, catalytic ions and polymer network are mingled together by covalent bonding. Additional relevance of nonlinearity in materials science has effectively began with systematic study of electrochemical processes [27, 92], designing of self-oscillating gels [93], stationary space–time structures like crystallizing patterns of snowflakes [94, 95]. The diffusion kinetics is certainly accompanied in such assembly phenomenon. Dissipation of energies is attributing from thermodynamically stable crystal patterns. Thorough deliberation of those self-assembly processes requires facts and data to suggest true mechanism for patterning characteristics. It is useful to obtain the energy values and entropy levels of growing and fully grown space–time structures. It has been suggested that nonlinear science evolved to be a matter-of-fact for tailoring the material's properties even at nanoscales.

1.16   Non-equilibrium Crystallization Phenomena

Pattern formation in chemically reacting systems is ubiquitous in nature, and the coupling between oscillations and self-organization in non-equilibrium conditions gives rise to a wealth of complex spatial patterns. An attempt has been performed to accumulate some well-characterized non-equilibrium crystallization phenomena, which exhibit the growth of some typical crystal morphologies as presented in Fig. 1.11. Classical examples include Liesegang systems or rhythmic spatial precipitation [87, 88]. They are formed by periodical precipitation of at least two substrates, where reaction–diffusion mechanism controls the spatial rhythmicity. The electrochemical reactions [27, 28], fractals growth [52], formation of dendrites [96, 97], growth of diffusion-limited aggregation (DLA)-like branched patterns [98, 99], spherulitic crystal patterns [100–104], etc., are the other representative examples of non-equilibrium crystallization phenomena. Apparent morphology formed during these crystallization processes when one product phase diffuses into another in periodic mannered. Similar to Liesegang systems, reaction–diffusion mechanism, and distances from thermodynamic equilibrium, surface energy has been analyzed which effectively play roles in controlling the crystal morphologies whether they emerged from solution state, colloidal phase, or phase separations. The typical examples of some common types of non-equilibrium crystal patterns have been illustrated in Fig. 1.11.

Chemical engineers are highly astonished through the formation of striking and fascinating   spatiotemporal   patterns   which   observed   especially   at   the non-equilibrium conditions in comparison with the material scientists. There are

Fig. 1.11 A relationship between the distance from equilibrium and morphology of growing crystal patterns (Adopted from Ref. [11])

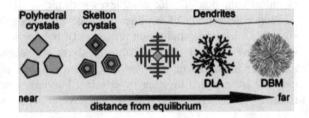

two approaches which possibly differentiate the systems in different ways: Firstly, the chemical engineer has focused on the molecules. This is emphasizing that, how a macroscopic structure is formed from molecular-level interactions. However, a pure material scientist is usually dependent on the functionality of thermodynamical states for the patterning behaviors. Some recent work was investigated in order to comprehensive design of crystal patterns by controlling the crystallites, sizes, and morphologies. Since many properties of a structural system are dependents on the macro- and microscopic crystallites. The manipulation and fabrication of crystal patterns at definite orders is significant for specific application and improvement in the technology. Various techniques have already been applied reasonably in modifying the crystal patterns and morphologies are starting from macroscopic levels.

In this section, three topical non-equilibrium crystallization phenomena have been presented and briefly described, namely (1) dendritic crystal patterns, (2) growth of DLA-like crystal patterns, and (3) spherulitic crystal patterns.

1.16.1 Dendritic Crystal Growth

Dendritic crystal growth is an example of spontaneous pattern generation in non-equilibrium systems. They are typically showing a tree-like branched microstructure which is exceptionally attractive for procedural and meticulous grounds. In contrast to multifarious biological systems, these types of patterns are attributing very common illustrations of self-organization [97–99] (Fig. 1.12).

Formation of dendritic crystal patterns has been investigated in variety of crystallization systems. Such patterns are usually obtained due to instability arises between interfaces of solid–liquid boundaries, having a hierarchy in branching orders which composed of primary, secondary and higher level of branches. Existing theory of dendritic crystal patterns are remained restricted in the estimate of steady-state characteristics, and also for growing directions. Till now, it has examined that commonly dendrites grow in cubic symmetric geometry correspond to the main crystal axes, or inside the hexagonal lattice. The dendritic crystal patterns grow in an oscillatory chemical system would be new study in this field.

A number of experimental and simulation work has been carried out in recent decades which obviously improved the elementary knowledge about the growth of dendritic crystal patterns. Some modern technique has also been has investigated which can replicate the true geometry of dendrites and validated experimentally both in space and time scales. This has led to elucidate new pattern formation mechanisms which broaden the scope of our understanding in dendritic pattern formation in diverse field of material science.

The growth of crystallization phenomenon sometimes forms nontrivial and beautiful structures. Some of the crystals are associated with polyhedral structure, but more complex dendrite/fractal and seaweed structures can also appear [105, 106]. Dendrites are anisotropic ramified structures with nearly paraboloidal growing

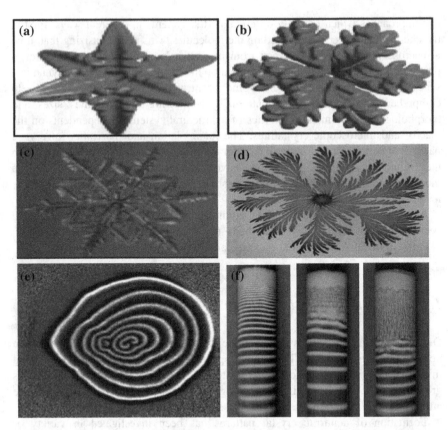

Fig. 1.12 Non-equilibrium crystallization patterns. **a–c** The dendritic crystal patterns, **d** DLA-like crystal patterns and **e–f** the Leisegang-type spatial patterns

tips. On the other hand, seaweed-like morphologies consist (in 2D) of doubloons that are double-finger-like structures separated by narrow channels [107] are more isotropic. Snow crystals [108] which are formed by the combination of dendritic and faceted growth is the best example.

There are the two ways by which solidification processes can precede further. The first process covers the crystallization of atoms which are transported to the crystal surface. When the crystal grows from a solution [109], this process becomes important under some fundamental basic condition like when the growth occurs in an under cooled melt, no chemical transport processes are required. The second process covers incorporation of the atoms into the crystal. Finally, the third process includes the transportation of latent heat away from the surface. The whole growth is caused by the slowest process [110].

When the nucleation process that starts at the surface of any solids (e.g., the organization of indissociable atom clusters on the surface of the solid) occurs with less probability, faceted crystals are formed, and therefore, the surface of the crystal is smooth on molecular level. This condition involves that the kinetic process is very slow, and the growth is controlled by the kinetics of incorporating new elements. Usually, Wulff construction apparent method is used for identifying the shape of a growing faceted crystal where the orientation-dependent growth rate should be used instead of the surface free energy of the respective planes [111].

Various kinds of other growth phenomena can also produced nontrivial patterns. The M-S mechanism in crystal growth involves destabilization of the growing surface which can also be caused by several other effects that enhance the growth of small protrusions of the surface. Here, we have discussed two types of growth processes, i.e., DLA and spherulitic.

1.16.2   Diffusion-Limited Aggregation-Like Crystal Growths

Diffusion-limited aggregation (DLA) is a theory by which researcher can usefully validates the structure–property relationship of very complex and intricate morphological patterns. In last two decades, the scientific community showed a great interest in DLA experiments, because of a quite resemblance of these morphologies with various phenomena of physical, chemical, and biological systems. It takes place in non-living systems (e.g., mineral deposition, dendritic crystal growth, snowflakes growth, lightening paths, electrochemical deposition, polymer crystallization in thin film), or living nature (e.g., coral formation, bacterial colony) [98].

The DLA model has been investigated in 1981, by Witten and Sander [112]. This was catching the attention of researchers due to property of various levels of crystal patterns especially in non-equilibrium state can be simulated instantly. The standard DLA model includes some basic terminologies which are conceptually helpful to describe complexity of the growth process.

Originally, the diffusion is a random motion of particles within a system. In DLA growth process, the motion of individual particles considered as random with respect to growth direction. Sometimes, it has been observed that particles walk extreme far comparative to their starting point. The average of walkable distance of all particles within a random walk (Brownian motion) is zero. While one particle moves into this direction, another moves into another direction. In diffusion motion, a net transport of particles has been estimated, if the solution is not uniform throughout. Contrast to the normal flow, the particles under investigation move more or less into the same direction. Diffusion mechanism allows interacting particles to others and permanently attached to form an aggregate structure. The driving energies are acting among the crystallizing entities may be feeble or high which depends upon the atomic nature of molecules. Since these molecules are certainly cover up with electronic charges which facilitate the emergence strong driving force. Thus, there will be possibility for again energy content when aggregates are formed. This energy may provide driving forces for aggregation, and also lowering down the surface energies. These factors can be useful to induce the stability of the growing structures.

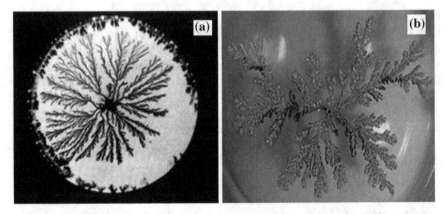

Fig. 1.13 Polypyrrole aggregate grown through diffusion-limited polymerization (**a**). A DLA cluster grown from a copper sulfate solution in an electrodeposition cell (**b**) [116]

Nature prefers symmetry. Mountains, snowflakes, branches of trees, shores of continent, and so forth are some visible examples of symmetric structures in nature. More patterns of self-similar structures encountered with mankind are cancer, piles, and so forth. Retinal circulation of the normal human retinal vasculature is, also, statistically self-similar and fractal. Hence, fractals are one of the most important topics in biology and medicinal field which generally covers the study of (a) the understanding of spatial shape and branching structure, and (b) the analysis of time varying signal. By knowing the branching structures of tissues and organs, biologists use this to discriminate between normal and pathological structures.

This has made the growth phenomenon of complex structure an interesting area of research for a long time [113–115]. Several models have been presented to understand the phenomenon out of which DLA model had received much attention as this is very common in nature [116]. Fractals are self-similar objects with non-integer dimension; they are also important to determine the macroscopic properties of the system by microscopic dynamics of system, which has been an area of scientific interest for a long time. Electrochemical deposition and some polymerization processes are the most well-known examples (Fig. 1.13).

1.16.3   Growth of Spherulitic Crystal Patterns

Spherulites are spherical crystal patterns, considered as a centrally nucleated and grown radially into fibrillar crystal organizations. Based on the chemical components (molecular geometry and molecular weight) and physical properties (crystallization conditions and crystalline lattice), the spherulites exhibit variety of shapes. The material systems under specific conditions can form spherulitic crystal

Fig. 1.14 Schemes for growth mechanisms of the spherulites

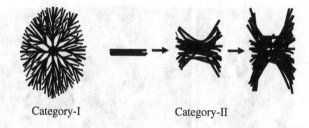

Category-I   Category-II

pattern of different morphology. Sometimes, a single material system may form two or more types of spherulitic crystal patterns [117], as presented in the Fig. 1.14.

Spherulitic crystal patterns have been investigated in various crystallizing processes of polymeric systems. It was also observed in solidification of protein molecules and some other biological functional materials. However, mechanism for the development of spherulitic patterns in a particular reacting system, as also the kinetics parameters associated with growing structures are not yet completely investigated even nowadays. The kinetics factors, rate of growth, unified growth model, elementary theories, and emergence of radial networks have attracted many attentions to the researchers. Efforts have been made to measure the surface morphology and geometry of spherulitic crystal patterns. But this is not the confirmations or indication for entire crystal perfection in the spherulitic crystal morphology. The surface analysis could be useful to deduce relationship between crystal perfection and morphological geometries. The morphological orders and geometry of various spherulitic forms are qualitatively estimated and validated by means of growth kinetics and analysis of surface energies. The mechanism for nucleation and theory for growth of such spherulitic patterns have also been proposed.

A major theory was used to describe that spherulitic patterns are formed by separation of impurities from crystalline phase during the growth process. It creates a boundary layer just beyond the growing fronts. Because of compositional differences of above-mentioned two phases, an instability is developed which facilitates fibrillar network in the regular spherulites [119, 120]. Goldenfeld [121] has also suggested that spherulitic patterns with non-crystallographic branching are accompanied by diffusion-mediated growth. The variation in the linear radius $R(t)$ is the function of the time t. This was experimentally verified in growth of diverse morphologies of spherulites especially in supersaturation conditions [119].

Early studies on spherulitic growth were sectioned in different subjects; because they have investigated in numbers of material systems of various characteristics. Some well-established morphologies of spherulite are presented in Fig. 1.15. Efforts have been made to propose a common feature of growth and mechanism of spherulitic patterns. However, it observed that mechanism of spherulitic growth is not depends only on the molecular property and chemical structures of material systems. The morphologies and physical characteristics of a spherulite from a system to other materials are varied drastically and found to be reliable on few key factors.

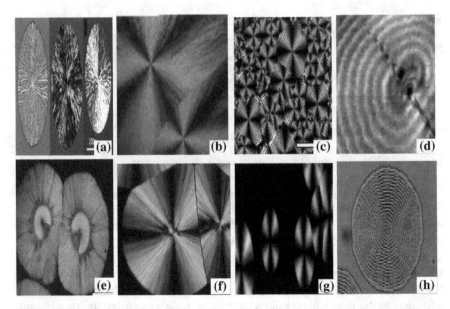

Fig. 1.15 Spherulites reported in **a** phase-field model. **b** Malonamide. **c** Polyethylene styrene **d** 4-Cyano-4-decycloxybiphenyl liquid crystals. **e** Polyethylene terephthalate **f** Polyesteramide **g** D-Sorbitol and **h** PVDF–PVAc system [118]

1.17 Significance of the Research Work

Oscillating chemical systems have advanced from rudimentary studies to a very popular subject of contemporary nonlinear science. The BZ reaction system has become a model chemical reaction which essentially exhibits a complex behavior often at far-equilibrium conditions. The early use of these reactions was restricted to physiochemical illustrations of various biological phenomena, such as morpho-genesis in biology, glycolytic oscillations, heartbeats, circadian rhythms, and periodic variations of hormone levels in animal body, and pattern formation on animal skin. It has been assumed that highly intricate structures well resembled with complex forms of life in nature grown over billions of years are common examples of self-organization and oscillations. A good understanding of dynamics and thermodynamical properties behind these biological oscillators would be imperative to our understanding of all living things.

By introduction of reaction–diffusion mechanism, self-organization, and self-regulation perceptions in the oscillating reactions, its applications have grown substantially in recent years. The reaction–diffusion mechanism is found to be very usual in diverse kinds of natural phenomenon that employed to assemble and fabricate the structures on the length scales. On the other hand, self-organization is treated as a fantastic phenomenon by which a spontaneous dissipative pattern could be possible by input of energy and matter in non-equilibrium conditions.

For their aesthetic appeal, richness of dynamic behaviors, theoretical challenges, and relevance to life, the applications of oscillatory reactions advance extensively in interdisciplinary research fields. The efforts have begun investigators to seek insights into the attractive space–time structures that occur in material processing, stability of growing body, and non-equilibrium crystallization processes. Under the non-equilibrium instability conditions, many of these physical processes give rise variety of spontaneous dissipative structures which could be effective to deduce the structure–property relationship for preparation of new materials of their tailor-made quality. The significance of nonlinear dynamics was initially demonstrated in material science. It has usefully shown that, how a polymerizing reaction is carried out by internal couplings of oscillatory reaction network and reaction–diffusion mechanism.

Other selected applications of nonlinear science include electrochemical processes, designing of self-oscillating gels, thermally stable crystallization patterns, crystal growth of protein molecules, biological membranes, and other aggregation phenomena. Functionality of a crystal pattern is largely depended on the molecular structures of material system; thus, fabrication of definite geometries is essential to set a desired application. It has also been implied that nonlinear dynamics come into views to be helpful for scheming novel objects even at nanoranges.

Non-equilibrium soft matter is an emerging research field, which could be able to explain a range of phenomenon of self-organization by association of reaction–diffusion mechanism and phase separations. Those phenomena are necessary for preparation of new materials of specific applications and can have possible relevance in the synthetic nanodevices. Thus, applications of the oscillatory chemical reaction continue to grow with rapid progress of nonlinear science.

1.18 Scope and Objectives of the Research Work

Patterning phenomena in oscillatory reaction systems are evolved as the most important subject of study from the invention of the first spatiotemporal chemical waves in the Fe-/Ce-catalyzed BZ reaction. Considerable advance is achieved in the last 20 years optimizing, scheming, simulating, and knowing the BZ phenomenology in various reaction conditions. Various nonlinear phenomena in those chemically reacting processes facilitate a simpler model to understand various phenomena of chemical, physical, and biology. The classic BZ reaction consists of malonic acid (MA) as initial substrate which has been modified by a large number of organic compounds to study the unusual patterning behaviors. However, the selection of new substrate for a modified BZ reaction must fulfill the chemical property as confined in MA. The MA usually exists in diacid-form, which is readily converted into its reactive enol-form in the solution state. The bromination and oxidation are two important mechanistic steps of the BZ reaction. These reaction steps performed frequently with central carbon atom of enol-form of MA.

MA   CHD

The MA in BZ reaction has an important drawback which is producing carbon dioxide gas bubbles in the oscillating reaction process. Some attempts have been made to develop gas-free versions of the BZ reactions. The diketonic compound such as cyclohexanedione (CHD) is a well-set example in which the gas-free oscillations and formation of spatial patterns have been examined. The acetyl acetone (AA) and ethyl acetoacetate (EAA) can also bear an analogue diketonic property. The potential oscillations in these two compounds have been reported in well-stirred medium. Though, the characteristics of an oscillating reaction in stirred reaction may differ from the unstirred reaction medium. Thus, two diketonic compounds, namely AA and EAA, are selected to extend the study of patterns formation and oscillatory behaviors in unstirred BZ conditions.

Beside this, the succinic acid (SA) and adipic acid (ADA) is an analogue diacid compound similar to MA. Due to low solubility, the SA and ADA alone is unable to produce any sort of oscillations and pattern formation in BZ reaction condition. A possibility for oscillation and pattern formation can be created if SA and ADA will combine with another organic substrate of known property. Thus, dual-substrate mode of BZ reaction has been adopted for present investigation. Also, the SA and ADA are important chemical components of various biochemical studies. Due to its crystalline property and constructive crystallizing habits, the crystallization phenomena might be expected in the domain of intermediates, catalysts, and some stable products of BZ reaction at widened time intervals.

AA   EAA   SA   ADA

Till date, the self-organization phenomenon which occurs spontaneously and nonlinear chemical dynamics far from thermodynamic equilibrium has been exhaustively explored by numerous scientists all across the world [122–129]. Such self-organization is a signature of living systems, where various small units combined to each other and are capable of exhibiting apparent and unique phenomena which cannot be shown by the individual units themselves. Such dynamical phenomena were investigated for comparably simple reaction systems—for instance, thin layer of certain autocatalytic reactions self-organizes traveling waves and rotating concentration patterns. These patterns are significant both from a fundamental as well as an applied perspective; fundamentally their formation dynamics is intriguing, and applicationally, we note that similar wave structures can cause irregular cardiac rhythms with potentially fatal consequences. Epstein group [125–127] investigated a correlation between morphological structure, chemical oscillation, and pattern formation in heterogeneous system such as a gel. They continue to be interested in developing new reactions that have particularly desirable features, e.g., producing specific types of patterns or being photosensitive. The mechanisms of reactions that display complex dynamical behavior were also reported. Epstein's group believes that these phenomena are to be of importance in a variety of pattern formation phenomena in living systems and they continue to study the dynamical origin of pattern formation, attempt to design new kinds of patterns, and investigate how introducing feedback can induce or alter pattern formation. Taylor's group [128, 129] has significant expertise in the experimental and theoretical investigation of the rates of reactions and mass transport in the physical sciences. One particular interest of the group is nonlinear chemical dynamics and the design and control of oscillations and patterns in chemical systems. The group has extensive experience in the construction of kinetic models of oscillatory chemical reactions to simulate the experimental results.

On the national front, Gopinathan and his coworkers [130, 131] are focused on experimental studies and modeling of nonlinear dynamical phenomena in chemistry, biochemistry, and physiology. Based on extensive data sets from EEG and ECG measurements, they claim that the human brain and heart follow chaotic dynamics. Recently, they have applied nonlinear theory to explain the experimentally observed oscillations of CO_2 gas over a layer of melting ice [130]. Das et al. [132, 133] investigated the growth of nanostructured fractal pattern and oscillations in potential during electropolymerization of pyrrole, mono- and mixed surfactant, and aniline. They found that the diameter of the fractal pattern was in the nanometer range and proposed a suitable mechanism for the fractal growth. They monitored oscillation in potential and were interrelated with fractal growth phenomenon.

Interests on the theoretical and experimental studies of oscillating chemical reactions are in the increasing trend and are now carried out at many laboratories although oscillating chemical reactions reached the status of a branch of chemistry laboratory only in late sixties, regarded as curiosities, influenced by the discovery of oscillations in biochemical pathways [134]. There is a long history [135] that has unfolded in this area paving the way for the explosion of activity of last three decades. The first significant event was made by Robert Boyle in seventeenth century. Boyle noted a periodic "flaring-up" of phosphorous in a loosely Stoppard flask arising from interaction of chemical kinetics and diffusion. The reaction between phosphorous and oxygen is a branched chain process that leads to an ignition. The ignition consumes oxygen available in the flask, and the reaction does not immediately recommence. Instead, the oxygen concentration must reach a critical value before the chain branching leads to another ignition process.

Other isolated observations were reported in a wide range of physical and chemical systems. In 1828s, Fechner [136] described an electrochemical cell producing an oscillating current; this was the first published report of oscillations in a chemical system. Ostwald [137] in 1899 observed that the rate of chromium dissolution in acid varied in a periodic form. Owing to the fact that both the systems were inhomogeneous, the existence of homogeneous oscillation reactions was believed to be impossible [138]. William C. Bray had first studied the reaction of iodate, iodine, and hydrogen peroxide, described as the first homogeneous isothermal chemical oscillator [7] at the University of California in 1921. During formation of oxygen and water through the decomposition of hydrogen peroxide, the evolution rate of oxygen and iodine concentration was found to vary periodically. A dark period was observed between 1925 to 1960 in the field of nonlinear dynamics because of the reason that theoretical chemists [139] held the view that oscillations in chemical systems were impossible and that these were an artifact of dust or bubbles. There was firm convection that second law of thermodynamics would just not allow the sort of abnormality—at least in homogeneous systems. Perhaps the greatest victim of that completely incorrect prejudice was Belousov [140], who made attempts to publish the observations of oscillations in the reaction that now bears his name jointly with Anatol Zhabotinsky. In 1920, a simple model had been developed by Lotka [141], based on two sequential autocatalytic reactions, that gave sustained oscillations. This model, though it does not apply to any real chemical systems, has provided considerable inspiration to ecologists.

Heilweil et al. [142] reported on sequential oscillations as dominant form of behavior, in mixed-substrate BZ system. They employed manganese sulfate as the catalyst and used three different substrates, MA, AA and EAA. They have monitored the oscillations by measuring the redox potential of the solution in between a platinum electrode and reference electrode as well as bromide concentration with a specific ion electrode, at 31 °C.

Srivastava et al. [143, 144] analyzed the dependence of the wavelength to the rhodium-catalyzed BZ oscillator in identifying the primary step of inhibition and reported the establishment of the response to the system in light irradiation. Dual-frequency oscillations had been reported by Srivastava et al. [145] in a number of systems containing single organic substrate, o-hydroxy acetophenone/m-hydroxy acetophenone/p-hydroxy-acetophenone with a number of catalysts Ce^{4+}, $Fe(phen)_3^{2+}$ and $Ru(bpy)_3^{2+}$ with acidic bromate. Such oscillations had been also reported with a few other organic compounds, AA, EAA and MA with acidic bromate. The dual-frequency oscillators seem to be very important for the systematic study of entertainment phenomena between oscillators coexisting in the system, competition among them, period doubling, and period multiplicity bifurcation are the factors which lead to the quasi-periodicity and chaos. Rastogi et al. [146] reported sequential oscillations and decay for ascorbic acid-cyclohexanone–Ce^{4+}–BrO_3^-–H_2SO_4 system in a batch reactor.

Reference [147] reports the simultaneous analysis of spatiotemporal pattern formation in electrolytic oxidation of sulfide on a platinum disk using electrochemical methods and a CCD camera. A series resistor and a large electrode area facilitated measurement of kinetic instability across a wide range of the potential curve. Spatial patterns such as pulses and labyrinthine stripes, as well as temporal patterns like fronts, twinkling eyes, and alternate deposition and dissolution were observed at different experimental conditions.

The formation of Turing patterns in the CDIMA reaction was performed experimentally in a spatial open gel disk reactor where all the input species was fed onto one side by a CSTR [148]. After investigating Turing pattern in CIMA reaction, Watzl and Munster investigated Turing-like spatial pattern in another chemical system consisting of polyacrylamide–methylene blue–sulfide–oxygen system [149]. In this system, they found a variety of spatial structures such as hexagons strips and zigzag patterns.

A variety of patterns was observed when carboxylic acid crystallizes from aqueous solution and mixed with agar or from alcoholic solutions [150]. Oxalic acid crystallizes in fractal-like morphology with fractal dimension D = 1.65. Succinic acid and adipic acid show tree-like and dendritic morphology, respectively.

Wang et al. [151] reported the growth of self-organized branched structure patterns of barium carbonate crystals. Their experimental results describe that these branched crystal formed through reaction–diffusion which governed pattern evolution and selection in many chemical and biological systems. The dendritic crystal growth patterns that typically grown along principal crystallographic axes and had the hierarchical structure had been attracting much attention from scientists for several centuries. Li et al. [152] reported that the ZnO dendritic nanostructure as a new member of the ZnO family could be successfully prepared on Cu substrates by electrochemical deposition in the solution of $ZnCl_2$ + CA at a temperature of 90 °C.

This chapter, in this context, has aimed at understanding the nonlinearity in the chemical systems involving BZ reaction and is believed to be a useful contribution even in understanding various biological growth phenomena, which further can aid to developing strategies useful in medical sciences.

The growth of nanostructured fractal pattern and oscillations in potential during electropolymerization of pyrrole, mono- and mixed surfactant, and aniline were reported by Das et al. [153, 154]. They found nanosized diameter of the fractal pattern, and a suitable mechanism for the fractal growth had been proposed. They monitored oscillation in various potentials and explained their interrelation with fractal growth phenomenon with proposed mechanism of the formed fractal pattern.

Many organic substrates have been used in place of MA in of BZ reactions. Kasperek and Bruice [155] listed a number of organic acids that did not generated oscillations in their hands, and the mechanism in these cases was apparently similar to that with MA. Some organic substrates lead to reactions so different from that with malonic acid that they were discussed as major modifications. Use of unsaturated dicarboxylic acids [156, 157] as the organic substrates had the advantage that no CO_2 was generated, which hinders the oscillations. Most of the interesting results had been obtained with ketones and diketones. Ketones such as cyclohexanone, butanone, and 3-pentanone gave oscillations in cerium-catalyzed systems [158]. One peculiarity of ketone substrates was that their bromo derivatives did not react with Ce (IV) to liberate Br^- and hence could not be explained by classic FKN mechanism. Nadeem et al. [159] had used ketones as cosubstrates with resorcinol as the main BZ substrate. Lone et al. [160] studied oscillatory behavior of BZ reaction with the mixed substrate systems containing the gallic acid and different methyl ketones (acetone, butanone, and pentanone) at different temperatures in sulfuric acid medium. An important aspect of the BZ organic substrates was the use of polyphenols and polyanilines. Orban and Körös had reported a good wealth of dynamic phenomena in polyphenols and polyanilines even in the absence of a metal-ion catalyst and proposed a bromide ion-controlled radical-based mechanism [161].

The oscillating chemical system with catechol as organic substrate in a batch reactor under unstirred conditions with Mn (II) as catalyst was reported [162]. These aromatic substrates were especially important in the study of spatial periodicities and shown a variety of spiral and scroll behaviors. Salter and Sheppard [163] reported a dual-frequency oscillator with EAA as substrate. Its behavior was much like the MA system except that a set of high-frequency, negatively damped oscillations in both redox potential and bromide ion were superimposed on early part of the induction period.

References

1. Epstein, I.R., Pojman, J.A.: Introduction to Nonlinear Chemical Dynamics: Waves, Patterns and Chaos. Oxford University Press, New York, Oscillations (1998)
2. Nicolis, G., Prigogine, I.: Self-organization in Nonequilibrium Systems: From Dissipative Structures to Order Through Fluctuations. Wiley, New York (1977)
3. Epstein, I.R., Showalter, K.: Nonlinear chemical dynamics: oscillations, patterns, and chaos. J. Phys. Chem. **100**, 13132 (1996)
4. Field, R.J., Burger, M.: Oscillations and Traveling Waves in Chemical system. Wiley-Interscience, New York (1985)
5. Turing, A.M.: The chemical basis of morphogenesis. Philos. Trans. R. Soc. London, Ser. B **237**, 37 (1992)
6. Kepper, P.D., Kustin, K., Epstein, I.R.: A systematically designed homogeneous oscillating reaction: the arsenite-iodate-chlorite system. J. Am. Chem. Soc. **103**, 2133 (1981)
7. Bartosz, A., Grzybowski, K.M., Bishop, C.J., Campbell, M.F., Stoyan, K.S.: Micro- and nanotechnology via reaction–diffusion. Soft Matter **1**, 114 (2005)
8. Ball, P.: The Self-Made Tapestry: Pattern Formation in Nature. Oxford University Press, New York (1999)
9. Fukami, K., Nakanishi, S., Yamasaki, H., Tada, T., Sonoda, K., Kamikawa, N., Tsuji, N., Sakaguchi, H., Nakato, Y.: General mechanism for synchronization of electrochemical oscillations and self-organized dendrite Electrodeposition of metals with ordered 2D and 3D microstructures. J. Phys. Chem. C **111**, 1150 (2007)
10. Yoshida, R., Sakai, T., Hara, Y., Maeda, S., Hashimoto, S., Suzuki, D., Murase, Y.: Self-oscillating gel as novel biomimetic materials. J. Cont. Releas. **140**, 186 (2009)
11. Oaki, Y., Imai, H.: Experimental demonstration for the morphological evolution of crystals grown in gel media. Cryst. Growth Des. **3**, 711 (2003)

12. Vanag, V.K., Epstein, I.R.: Pattern formation in a tunable medium: the Belousov-Zhabotinsky reaction in an aerosol-OT microemulsion. Phys. Rev. Lett. **87**, 228301 (2001)

13. Rastogi, R.P., Srivastava, R.C.: Interface-mediated oscillatory phenomena. Adv. Colloid Interface Sci. **93**, 1 (2001)

14. Vanag, V.K., Epstein, I.R.: Dash waves in a reaction-diffusion system. Phys. Rev. Lett. **90**, 098301 (2003)

15. Jia, L., Yu, S., Qiang, C., Qianyao, S., Hengde, L., Xihua, C., Xiaoping, W., Yunjie, Y., Vrieling, E.G.: Patterning of nanostructured cuprous oxide by surfactant-assisted electrochemical deposition. Cryst. Growth Des. **8**, 2652 (2008)

16. Atkin, A., Ross, J.: Statistical construction of chemical reaction mechanisms from measured time-series. J. Phys. Chem. **99**, 970 (1995)

17. Rastogi, R.P., Srivastava, R.C.: Casuality principle, non-equilibrium thermodynamics and non-linear science of open systems. J. Sci. Ind. Res. **67**, 747 (2008)

18. Cross, M.C., Hohenberg, P.C.: Pattern-formation outside of equilibrium. Rev. Mod. Phys. **65**, 851 (1993)

19. Bray, W.C.: A periodic reaction in homogeneous solution and its relation to catalysis. J. Am. Chem. Soc. **43**, 1262 (1921)

20. Bray, W.C., Liebhafsky, H.A.: Reactions involving hydrogen peroxide, iodine and iodate ion. I. Introduction. J. Phys. Chem. **53**, 38 (1931)

21. Shaw, D.M., Pritchard, M.O.: The existence of homogeneous oscillating reactions. J. Phys. Chem. **72**, 1403 (1968)

22. Weiser, H.B., Garrison, A.: The oxidation and luminescence of phosphorus. J. Phys. Chem. **25**, 61 (1921)

23. Rayleigh, L.: A study of the glow of phosphorus. Periodic luminosity and action of inhibiting substances. Proc. Roy. Soc. Ser. A **99**, 372 (1921)

24. Douglas, J.M., Rippon, D.T.: Unsteady state process operation. Chem. Eng. Sci. **21**, 305 (1966)

25. Nicolis, G., Portnow, J.: Chemical oscillations. Chem. Rev. **73**, 365 (1973)

26. Bush, S.F.: Vapor phase reaction of methyl chloride. Proc. Roy. Soc. Ser. A **309**, 1 (1964)

27. Das, I., Mishra, S.S.: Fractal growth and oscillation during electrochemical deposition in Pb-Zn binary system. Indian J. Chem. Sect. A **39**, 1005 (2000)

28. Suter, L.M., Wong, P.: Nonlinear oscillations in electrochemical growth of Zn dendrites. Phys. Rev. B **39**, 4536 (1989)

29. Lotka, A.J.: Undamped oscillations derived from the law of mass action. J. Am. Chem. Soc. **42**, 1595 (1920)

30. Tyson, J.J.: Some further studies of nonlinear oscillations in chemical systems. J. Chem. Phys. **58**, 3919 (1973)

31. Field, R.J., Noyes, R.M.: Oscillations in chemical systems. J. Chem. Phys. **60**, 1877 (1974)

32. Winfree, A.T.: The prehistory of the Belousov-Zhabotinsky oscillator. J. Chem. Edu. **61**, 661 (1984)

33. Belousov B.P., Field R.J., Burger M.: A Periodic Reaction and Its Mechanism: in Oscillations and Traveling Waves in Chemical Systems Ed. Wiley, New York, pp. 605–613 (1985)

34. Hall, L.D., Waterton, J.C.: A method for determining the spatial distribution of spin-labeled organic ligands covalently bound to a noncrystalline surface: dipolar contribution to nitroxide EPR spectrum. J. Am. Chem. Soc. **101**, 3697 (1979)

35. Atkins, P., de Atkin, P.J.: Physical Chemistry. Oxford University Press, pp. 60–80 (2006)

36. Zaikin, A.N., Zhabotinsky, A.M.: Concentration wave propagation in two dimensional liquid-phase self-oscillating system. Nature **225**, 535 (1970)

37. Demas, H.N., Diemente, D.: An oscillating chemical reaction with a luminescent indicator. J. Chem. Edu. **50**, 357 (1973)

38. Zhabotinsky, A.M.: A history of chemical oscillations and waves. Chaos. **1**, 379 (1991)

39. Noyes, R.M., Field, R.J., Thompson, R.C.: Mechanism of the reduction of the Br(V) by single electron reducing agents. J. Am. Chem. Soc. **93**, 7315 (1971)
40. Degn, H.: Effect of bromine derivatives of malonic acid on the oscillating reaction of malonic acid, cerium ions and bromated. Nature **213**, 589 (1967)
41. Wood, P.M., Ross, J.: A quantitative study of chemical waves in the Belousov- Zhabotinsky reaction. J. Chem. Phys. **82**, 1924 (1985)
42. Field, R.J., Burger, M.: Oscillations and Traveling Waves in Chemical Systems. Wiley, New York, pp. 120–140 (1985)
43. Noyes, R.M., Field, R.J., Körös, E.: Oscillations in chemical system I. Detailed mechanism in a system showing temporal oscillations. J. Am. Chem. Soc. **94**, 1394 (1972)
44. William, C.T.: A threshold phenomenon in the Field-Noyes model of the Belousov-Zhabotinsky reaction. J. Math. Anal. Appl. **58**, 233 (1977)
45. Zaikin, A.N., Zhabotinsky, A.M.: Autowave processes in distributed chemical system. J. Theor. Biol. **40**, 45 (1973)
46. Scott, S.K.: Oscillations, Waves, and Chaos in Chemical Kinetics. Oxford University Press, New York (1994)
47. Epstein, I.R., Showalter, K.: Nonlinear chemical dynamics: Oscillations, patterns and chaos. J. Phys. Chem. **100**, 13132 (1996)
48. Zaikin, A.N., Zhabotinsky, A.M.: Concentration wave propagation in two-dimensional liquid-phase self-oscillating system. Nature **225**, 535 (1970)
49. Winfree, A.T.: The prehistory of the Belousov-Zhabotinsky oscillator. J. Chem. Educ. **61**, 661 (1984)
50. Zhabotinsky, A.M.: A history of chemical oscillations and waves. Chaos **1**, 379 (1991)
51. Epstein, I.R.: The role of flow systems in far-from-equilibrium dynamics. J. Chem. Educ. **66**, 191 (1989)
52. Zhabotinsky, A.M.: Periodical oxidation of malonic acid in solution: a study of the Belousov reaction kinetics. Biofizika **9**, 1306 (1964)
53. Field, R.J., Koros, E., Noyes, R.M.: Oscillations in chemical systems: Thorough analysis of temporal oscillation in bromate-cerium-malonic acid system. J. Am. Chem. Soc. **94**, 8649 (1972)
54. Vidal, C., Roux, J.C., Rossi, A.: Quantitative measurement of intermediate species in sustained Belousov-Zhabotinskii oscillations. J. Am. Chem. Soc. **102**, 1241 (1980)
55. Hansen, E.W., Ruoff, P.: Determination of enolization rates and overall stoichiometry from proton NMR records of the methylmalonic acid Belousov-Zhabotinskii reaction. J. Phys. Chem. **93**, 2696 (1989)
56. Balcon, B.J., Carpenter, T.A., Hall, L.D.: Methacrylic acid polymerization. Traveling waves observed by nuclear magnetic resonance imaging. Macromolecules **25**, 6818 (1992)
57. Roelofs, M.G., Jensen, J.H.: EPR oscillations during oxidation of benzaldehyde. J. Phys. Chem. **91**, 3380 (1987)
58. Jimenez-Prieto, R., Silva, M., Perez-Bendito, D.: Approaching the use of oscillating reactions for analytical monitoring. Analyst **123**, 1 (1998)
59. Ball, P.: The Self-Made Tapestry, Pattern formation in nature. Oxford University Press, New York, pp. 1–287 (1999)
60. Painter, K.J., Hunt, G.S., Wells, K.L., Johansson, J.A., Headon, D.J.: Towards an integrated experimental-theoretical approach for assessing the mechanistic basis of hair and feather morphogenesis. Interface Focus **2**, 433 (2012)
61. Scott, S.K.: Oscillations, Waves, and Chaos in Chemical Kinetics. Oxford University Press, New York (1994)
62. Turing, A.M.: The chemical basis of morphogenesis. Philos. Trans. R. Soc. Lond. Ser. B **237**, 37 (1992)
63. Maselko, J., Reckley, J.S., Showalter, K.: Regular and irregular spatial patterns in an immobilized-catlyst Belousov-Zhabotinsky reaction. J. Phys. Chem. **93**, 2774 (1989)
64. Winfree, A.T.: Scroll-shaped waves of chemical activity in three dimensions. Science **181**, 937 (1973)

65. Tyson, J.J., Glass, L., Arthur, T.: Winfree (1942–2002). J. Theor. Bio. **230**, 433 (2004)
66. Biosa, G., Bastianoni, S., Rustici, M.: Chemical waves. Chem. Eur. J. **12**, 3430 (2006)
67. Castets, V., Dulos, E., Boissonade, J., De Kepper, P.: Experimental evidence of a sustained standing Turing-type nonequilibrium chemical pattern. Phys. Rev. Lett. **64**, 2953 (1990)
68. Ouyang, Q., Swinney, H.L.: Transition from a uniform state to hexagonal and striped Turing patterns. Nature **352**, 610 (1991)
69. Vang, V.K., Epstein, I.R.: Pattern formation mechanism in reaction-diffusion systems. Int. J. Dev. Biol. **53**, 673 (2009)
70. Hoar, T.P., Schulman, J.H.: Transparent water-in-oil dispersions: the oleopathic hydro-micelle. Nature **152**, 102 (1943)
71. Bourrel, M., Schechter, R.S.: Microemulsions and Related Systems—Formulation, Solvency, and Physical Properties. Marcel Dekker, Inc., New York, pp. 30–78 (1988)
72. Schwartz, L.J., DeCiantis, C.L., Chapman, S., Kelley, B.K., Hornak, J.P.: Motions of water, decane, and AOT in reverse micelle solutions. Langmuir **15**, 5461 (1999)
73. Paul, B.K., Moulik, S.P.: Use and applications of microemulsions. Curr. Scie. **80**, 990 (2001)
74. Prince, L.M.: Microemulsions: Theory and Practice. Academic Press, New York, pp. 111–129 (1977)
75. Balasubramanian, D., Rodley, G.A.: Incorporation of chemical oscillators into organized surfactant assemblies. J. Phys. Chem. **92**, 5995 (1988)
76. Vanag, V.K.: Waves and patterns in reaction–diffusion system. Belousov-Zhabotinsky reaction in water-oil-microemulsions. Phys. Usp. **47**, 923 (2004)
77. Hildebrand, M.: Self-organized nanostructures in surface chemical reactions: Mechanisms and mesoscopic modeling. Chaos **12**, 144 (2002)
78. Sachs, C., Hildebrand, M., Voelkening, S., Wintterlin, J., Ertl, G.: Spatiotemporal self-organization in a surface reaction: From the atomic to the mesoscopic scale. Science **293**, 1635 (2001)
79. Shibata, T., Mikhailov, A.S.: Nonequilibrium self-organization phenomena in active Langmuir monolayers. Chaos **16**, 37108 (2006)
80. Epstein, I.R., Pojman, J.A., Steinbock, O.: Introduction: Self-Organization in Nonequilibrium Chemical Systems. Chaos **16**, 37101 (2006)
81. Stupp, S.I., LeBonheur, V., Walker, K., Li, L.S., Huggins, K.E., Keser, M., Amstutz, A.: Supramolecular materials: Self-organized nanostructures. Science **276**, 384 (1997)
82. Shenhar, R., Norsten, T.B., Rotello, V.M.: Polymer-mediated nanoparticle assembly: Structural, control and applications. Adv. Mater. **17**, 657 (2005)
83. Lopes, W.A., Jaeger, H.M.: Hierarchical self-assembly of metal nanostructures on diblock copolymer scaffolds. Nature **414**, 735 (2001)
84. Ray, W.H., Villa, C.M.: Nonlinear dynamics found in polymerization processes. Chem. Eng. Sci. **55**, 275 (2000)
85. Pojman, J.A., Trang-Cong-Miyata, Q. (eds.): Nonlinear dynamics in polymeric systems. ACS Symposium Series no. 869. Oxford University Press, New York, pp. 161–188 (2000)
86. Kawczyński, A.L.: Chemical reactions—from equilibrium, through dissipative structures to chaos, pp. 189–210. WNT, Warsaw (in Polish) (1990)
87. Okabe, Y., Kyu, T., Saito, H., Inoue, T.: Spiral crystal growth in blends of poly(vinylidene fluoride) and poly(vinyl acetate). Macromolecular **31**, 5823 (1998)
88. Ferreiro, V., Douglas, J.F., Warren, J., Karim, A.: Growth pulsations in symmetric dendritic crystallization in thin polymer blend films. Phys. Rev. E **65**, 51606 (2002)
89. Orlik, M.: Self-organization in nonlinear dynamical systems and its relation to the materials science. J. Solid State Electrochem. **13**, 245 (2009)
90. Epstein, I.R., Pojman, J.A.: Nonlinear dynamics related to polymeric systems. Chaos **9**, 255 (1999)
91. Cabral, J.T., Hudson, S.D., Harrison, C., Douglas, J.F.: Frontal photopolymerization for microfluidic applications. Langmuir **20**, 10020 (2004)
92. Sawada, Y., Dougherty, A., Gollub, J.P.: Dendritic and fractal patterns in electrolytic metal deposits. Phys. Rev. Lett. **56**, 1260 (1986)

93. Yoshida, R., Takahashi, T., Yamaguchi, T., Ichijo, H.: Self-oscillating gels. Adv. Mater. **9**, 175 (1997)
94. Libbrecht, K.G.: The physics of snow crystals. Rep. Prog. Phys. **68**, 855 (2003)
95. Henry, A.I., Courty, A., Goubet, N., Pileni, M.P.: How do self-ordered silver nanocrystals influence their growth into triangular single crystals. J. Phys. Chem. C **112**, 48 (2008)
96. Langer, J.S.: Instabilities and pattern formation in crystal growth. Rev. Mod. Phys. **52**, 1 (1980)
97. Doughterty, A., Kaplan, P.D., Gollub, J.P.: Development of sidebranching in dendritic crystal growth. Phys. Rev. Lett. **58**, 652 (1987)
98. Ferreira Jr, S.C.: Effects of the screening breakdown in the diffusion-limited aggregation model. Eur. Phys. J. **42**, 263 (2004)
99. Bogoyavlensky, V.A., Che-Rnova, N.A.: Diffusion-limited aggregation: A relationship between surface thermodynamics and crystal morphology. Phys. Rev. E **61**, 1629 (2000)
100. Magill, J.H.: Review spherulites: a personal perspective. J. Mater. Sci. **36**, 3143 (2001)
101. Goldenfeld, N.: Theory of spherulitic crystallization. J. Cryst. Growth **84**, 601 (1987)
102. Ryschenkow, G., Faivre, G.: Bulk crystallization of liquid selenium Primary nucleation, growth kinetics and modes of crystallization. J. Cryst. Growth **87**, 221 (1988)
103. Phillips, P.J.: Spherulitic crystallization in macromolecules, edited by D.T.J. Hurle, Handbook of Crystal Growth, Vol. 2, Elsevier, Amsterdam (1993)
104. Keith, H.D., Padden Jr, P.D.: A discussion of spherulitic crystallization and spherulitic morphology in high polymers. Polymer **27**, 1463 (1986)
105. Pimpinelli, A., Villain, J.: Physics of Crystal Growth. Cambridge University Press, Cambridge, pp. 131–167 (1998)
106. Glicksman, M.E., Marsh, S.P.: The dendrite. In: Hurle, D.T.J. (ed.) Handbook of Crystal Growth, vol. 1, pp. 1075–1121. North-Holland, Amsterdam (1993)
107. Fleury, V., Gouyet, J.-F., Leonetti, M. (eds.): Branching in Nature. Springer, Berlin, pp. 98–135 (2001)
108. Libbrecht, K.G.: Morphogenesis on Ice: The Physics of Snow Crystals. http://pr.caltech.edu/periodicals/EandS/archives/LXIV1.html
109. Ohara, M., Reid, R.C.: Modeling Crystal Growth Rates from Solution. Prentice-Hall, Englewood Cliffs, New Jersey, pp. 121–142 (1973)
110. Saito, Y.: Statistical Physics of Crystal Growth. World Scientific, Singapore, pp. 89–134 (1996)
111. Liu, X.-Y., Bennema, P.: Theoretical consideration of the growth morphology of crystals. Phys. Rev. B **53**, 2314 (1996)
112. Witten, T., Sander, L.: Diffusion-limited aggregation, a kinematic critical phenomenon. Phys. Rev. Lett. **47**, 1400–1403 (1981)
113. Jullien, R.: Aggregation phenomena and fractal aggregates. Contemp. Phys. **28**, 477 (1987)
114. Meakin, P.: The growth of rough surfaces and interfaces. Phys. Rep. **235**, 189 (1993)
115. Marsili, M., Maritan, A., Toigo, F., Banavar, J.R.: Stochastic growth equations and reparametrization invariance. Rev. Mod. Phys. **68**, 963 (1996)
116. Nazzarro, M., Nieto, F., Ramirez-Pastor, A.J.: Influence of surface heterogeneities on the formation of diffusion-limited-aggregation. Surf. Sci. **497**, 275 (2002)
117. Magill, J.H.: Review Spherulites: A personal perspective. J. Mater. Sci. **36**, 3143 (2001)
118. Phillips, P.J.: Spherulitic crystallization in macromolecules. In: Hurle, D.T.J. (ed.) Handbook of Crystal Growth, vol. 2. Elsevier, Amsterdam (1993)
119. Keith, H.D., Padden Jr, P.D.: A discussion of spherulitic crystallization and spherulitic morphology in high polymers. Polymer **27**, 1463 (1986)
120. Tracy S.L., Williams D.A., Jennings H.M.: The growth of calcite spherulites from solution II. Kinetics of formation, J. Crys. Growth **193**, 382 (1998)
121. Goldenfeld, N., Chan P.Y., Veysey J.: Dynamics of precipitation formation at geothermal hot springs, Phys. Rev. Lett. **96**, 254501 (2006)
122. Makki, R., Roszol, L., Pagano, J., Steinbock, O.: Tubular precipitation structures: materials synthesis under nonequilibrium conditions. Phil. Trans. R. Soc. A **370**, 2848 (2012)

123. Epstein, I.R., Pojman, J.A., Steinbock, O.: Self-organization in nonequilibrium chemical systems: a brief introduction. Chaos **16**, 037101 (2006)
124. Makki, R., Steinbock, O.: Nonequilibrium synthesis of silica-supported magnetite tubes and mechanical control of their magnetic properties. J. Ame. Chem. Soc. **134**, 15519 (2012)
125. Zhang, Y., Li, N., Gao, Y., Kuang, Y., Fraden, S., Epstein, I.R., Xu, B.: Post- self-assembly cross-linking of molecular nanofibers for oscillatory hydrogels. Langmuir **28**, 3063 (2012)
126. Rossi, F., Vanag, V.K., Epstein, I.R.: Pentanary cross-diffusion in water-in-oil microemulsions loaded with two components of the belousov–zhabotinsky reaction. Chem.–Eur. J. **17**, 2138 (2011)
127. Mao, S., Gao, Q.Y., Wang, H., Zheng, J., Epstein, I.R.: Oscillations and mechanistic analysis of the chlorite-sulfide reaction in a continuous-flow stirred tank reactor. J. Phys. Chem. A **113**, 1231 (2009)
128. Wrobel, M.M., Bánsági Jr, T., Scott, S.K., Taylor, A.F., Bounds, C.O., Carranzo, A., Pojman, J.A.: pH wave-front propagation in the urea-urease reaction. Biophys. J. **103**, 610 (2012)
129. Tinsley, M.R., Taylor, A.F., Huang, Z., Showalter, K.: Complex organizing centers and spatiotemporal behavior in groups of oscillatory particles. Phys. Chem. Chem. Phys. **13**, 17802 (2011)
130. Usharani, S., Srivdhya, J., Gopinathan, M.S., Pradeep, T.: Concentration of CO_2 over melting ice oscillates. Phys. Rev. Lett. **93**, 5634 (2004)
131. Sriram, K., Gopinathan, M.S.: A two variable dealy model for the circadian rhythm of Neurospora crassa. J. Theor. Biol. **231**, 23 (2004)
132. Das, I., Agrawal, N.R., Gupta, S.K., Gupta, S.K., Rastogi, R.P.: Fractal growth kinetic and electric potential oscillations during electropolymerization of pyrrole. J. Phys. Chem. A **113**, 5296 (2009)
133. Das, I., Goel, N., Agrawal, N.R., Gupta, S.K.: Growth pattern of dendrimers and electric potential oscillations during electropolymerization of pyrrole using monomono and mixed surfactants. J. Phys. Chem. B **114**, 12888 (2010)
134. Frank, G.M. (ed.):. Oscillatory Processes in Biological and Chemical Systems. Nauka, Moscow, vol. 1, pp. 132–187 (1967)
135. Zhabotinsky, A.M.: A history of chemical oscillations and waves. Chaos **1**, 379 (1991)
136. Fechner, GTh: Time series in the electrochemical oscillatory regime. J. Schweigg **53**, 61 (1828)
137. Ostwald, W.: Periodisch veraenderliche reaktionsgeschwindigkeite. Phys. Zeitsch. **8**, 87 (1899)
138. Deng, H.: Oscillating chemical reactions in homogenous phase. J. Chem. Ed. **49**, 302 (1972)
139. Rice, F.O., Reiff, O.M.: The thermal decomposition of hydrogen peroxide. J. Phys. Chem. **31**, 1352 (1927)
140. Winfree, A.T.: The prehistory of Belousov-Zhabotinsky oscillator. J. Chem. Educ. **61**, 661 (1984)
141. Lotka, A.J.: Undammed oscillations derived from the law of mass action. J. Ame. Chem. Soc. **42**, 1595 (1920)
142. Heilweil, E.J., Henchman, N.J., Epstein, I.R.: Sequential oscillations in mixed substrate Belousov-Zhabotinskii systems. J. Ame. Chem. Soc. **101**, 3698 (1979)
143. Srivastava, P.K., Mori, Y., Hanazaki, I.: Wavelength-dependent photo-inhibition of chemical oscillators: uncatalyzed oscillators with phenol and aniline as substrate. Chem. Phys. Lett. **177**, 213 (1991)
144. Srivastava, P.K., Mori, Y., Hanazaki, I.: Photo-inhibition of chemical oscillation in Ru(bpy)2+-catalyzed Belousov-Zhabotinskii reaction. Chem. Phys. Lett. **190**, 279 (1992)
145. Srivastava, P.K., Mari, Y., Hanazaki, I.: Duel frequency chemical oscillators with acetylophenols as substrates. J. Phys. Chem. **95**, 1636 (1991)
146. Rastogi, R.P., Mishra, G.P., Das, I., Sharma, A.: Sequential oscillations in bromine hydrolysis controlled oscillators in a closed reactor. J. Phys. Chem. **97**, 2571 (1993)

147. Zhao, Y., Wang, S., Varela, H., Gao, Q., Hu, X., Yang, J., Epstein, I.R.: Spatiotemporal pattern formation in the oscillatory electro-oxidation of sulfide on platinum disk. J. Phys. Chem. C **115**, 12965 (2011)

148. Rudovics, B., Barillot, E., Davies, P.W., Dulos, E., Boissonade, J., De, P.: Kepper, Experimental studies and quantitative modeling of Turing patterns in the (Chlorine dioxide, iodine malonic acid) reaction. J. Phys. Chem. A **103**, 1790 (1999)

149. Watzl, M., Munster, A.F.: Turing-like spatial patterns in a polyacrylamidemethylene blue-sulfide-oxygen system. Chem. Phys. Lett. **242**, 273 (1995)

150. Das, I., Kumar, A., Agrawal, N.R., Lall, R.S.: Non-equilibrium growth patterns of carboxylic acids crystallized on microslides. Ind. J. Chem. A **38**, 307 (1999)

151. Wang, T.: An-Wu Xu and H. Cölfen, Formation of self-organized dynamic structure patterns of barium carbonate crystals in polymer controlled crystallization. Angew. Chem. Int. Ed. **45**, 4451 (2006)

152. Li, Gao-Ren, Xi-Hong, Lu, Dun-Lin, Qu, Yao, Chen-Zhong, Zheng, Fu-lin, Qiong, Bu, Dawa, Ci-Ren, Tong, Ye-Xiang: Electrochemical growth and control of ZnO dendritic structures. Electrochim. Acta **50**, 5050 (2005)

153. Das, I., Choudhary, R., Gupta, S.K., Agrawal, P.: Nanostructured growth patterns and chaotic oscillations in potential during electropolymerization of aniline in the presence of surfactants. Phys. Chem. B **115**, 8724 (2011)

154. Das, I., Agrawal, N.R., Choudhary, R., Gupta, S.K.: Fractal growth patterns and oscillations in potential during electropolymerization of aniline with mono- and mixed surfactants. Fractals **19**, 317 (2011)

155. Kasperek, G.J., Bruice, T.C.: Observation on an oscillating reaction. The reaction of potassium bromate, ceric sulfate and a dicarboxylic acid. Inor. Chem. **10**, 382 (1971)

156. Beck, M.T., Varadi, Z.B.: Unsaturated dicarboxylic acids as substrates in oscillating reactions involving bromated sulfuric acid and a catalyst. React. Kinet. Catal. Lett. **6**, 275 (1977)

157. Showalter, K.: Pattern formation in ferroin-bromate system. J. Chem. Phys. **73**, 3735 (1980)

158. Farage, V.J., Stroot, P.H., Janjic, D.: Reaction chimiques oscillantes (type Belousov-Zhaboyinskii) impliquant des cetones cycliques et aliphatiques. Helv. Chim. Acta **60**, 231 (1977)

159. Ganaie, N.B., Nath, M.A., Peerzada, G.M.: Effect of mixed methyl ketones on the catalyzed resorcinol based oscillatory reaction at different temperatures. J. Indust. Eng. Chem. **16**, 634 (2010)

160. Lone, M.A., Nath, M.A., Ganie, N.B., Peerzada, G.M.: Oscillating behavior of galic acid-methyl ketone system catalyzed by metal ion. Ind. J. Chem. Sec. A **47**, 705 (2008)

161. Orban, M., Körös, E., Noyes, R.M.: Chemical oscillations during the uncatalyzed reaction of aromatic compounds with bromated. 2. A plausible skeleton mechanism. J. Phys. Chem. **83**, 3056 (1979)

162. Shah, I.A., Peerzada, G.M., Bashir, N.: A kinetic study on catechol-based Belousov-Zhabotinsky reaction. Inte. J. Chem. Kine. **45**, 141 (2013)

163. Salter, L.F., Sheppard, J.G.: A duel frequency Belousov-Zhabotinskii oscillating reaction with ethylacetoaceatate as organic substrate. Int. J. Chem. Kinett **14**, 815 (1982)

Chapter 2
Growth and Formation of Diffusion-Limited-Aggregation Crystal Pattern

2.1 Introduction

In nature, so many things around us have been created and designed beautifully but still most of them have appeared spontaneously such as living beings or snowflakes. The mechanism of pattern formation in nature, which is apparently produced via spontaneous processes that generate complex shapes and well-ordered structures, usually lies in the range of molecular as well as micro/nano scale. Some examples of nature which prefer symmetry like mountains, snowflakes, branches of trees, shores of continent, and so forth are some eye-catching examples of symmetrical structures in nature. More patterns of self-similar structures encountered by mankind are cancer, piles, and so forth. Retinal circulation of the normal human retinal vasculature is, also, statistically self-similar and fractal. Hence fractals are one of the most important topics in biology and medical fields, which generally cover the study of (a) the understanding of spatial shape and branching structure, and (b) the analysis of time varying signal. By knowing the branching structures of tissues and organs, biologists use this to discriminate between normal and pathological structures.

In 1981, Witten and Sander [1] proposed that diffusion-limited aggregation (DLA) describes a rule-based process which has been used to model many physical, biological, and social phenomena. According to the proposed theory in two dimensions it is easy to explain, a "particle" is placed randomly in the plane and undergoes a random walk until either it encounters an existing structure—initially a fixed random particle or *seed*—in which case it adheres, or its time limit expires and dies. The dendritic growth that results from the release, over time, of many particles has seen widespread application. Among the dendrite structure DLA has been used to model electrodeposition [2], urban cluster growth [3], root system growth [4], and even aspects of string theory [5]. There is much interest in DLA structure formation due to its importance and its interconnection with mathematics by its

© The Author(s) 2016

R. Srivastava et al., *Growth and Form of Self-organized Branched Crystal Pattern in Nonlinear Chemical System*, SpringerBriefs in Molecular Science, DOI 10.1007/978-981-10-0864-1_2

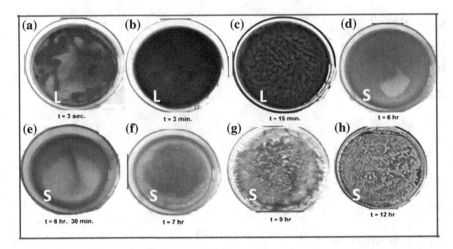

Fig. 2.1 Growth pathways of DLA pattern in ADA/EAA/Ce^{+4}/[Fe(Phen)$_3$]$^{+2}$/BrO$_3$/H$_2$SO$_4$ system. *Composition*   [ADA] = 0.0461 M,   [EAA] = 0.1212 M,   [Ce^{+4}] = 0.00486 M, [Fe(Phen)$_3$]$^{+2}$ = 0.0038 M, [BrO$_3$] = 0.0553 M, [H$_2$SO$_4$] = 0.42 M at 30 °C Petri dish (i.d) = 9.1 cm. The caption L and S denotes liquid phase and solid phase, respectively

fractal-like nature [4], with significant attempt having been devoted to measuring the fractal dimension of DLA formations [6].

Initially a Petri dish is used to keep an aliquot of homogenous red-color reaction solution which abruptly transforms into a blue-color solution in form of wave pattern concluding within 3 s (Fig. 2.1a). Blue color sustains for 3 min, and then the solution returns to light red with the second pattern formation (stationary-type pattern Fig. 2.1b). At certain phase of pattern development, chemical waves initiate to propagate from the light pacemaker centers in mosaic part of the pattern. The pattern remains stationary for about 75 min, and then fades away followed by two or three more oscillations, resulting in blue-color solutions stabilized for few seconds. After paused for few minutes, it takes a thick foggy brain-like structure, which after few minutes was found to transform into another red-sponge-like pattern (Fig. 2.1c) by a spontaneous branching process. The as-formed pattern was found to last for about 6 h before it disappeared and turned into a heterogeneous colloidal phase. The nucleation in solid phase started after 6 h duration from the initiation of the reaction, which continued to 30 min further, resulting in the formation of colloidal state at 7 h. It is clear that solid-phase nucleation occurs after 5 h and the final nanostructured DLA crystal formed after 9 h, which was maintained till its completely growth up to 12 h (Fig. 2.1d–h). This state of colloidal solution is stabilized for several hours that transforms into DLA crystal pattern which is stable for several days. These patterns form due to the coupling of diffusion process and nonlinear kinetics. However, as observed from the initial hexagonal structures formed in reaction system, surface tension of the solution is probably one of the determining factors. As surface tension is dependent on both the temperature and the concentration of the surfactants, one cannot rule out its role in explaining the phenomena observed.

The exact roles of diffusion process, nonlinear kinetics, and surface tension require further in-depth analysis, which is beyond the scope of the present paper. The wave front is quite sharp as illustrated in Fig. 2.1.

2.2 Colloidal State and Solid-Phase Nucleation

In the present case, a colloidal phase starts forming 7 h after the initiation of the reaction, which consists of nanoparticles suspended in the solution matrix. As a result of the self-aggregation of colloidal particles, solid-phase nucleation centers form with dendritic structure. These colloidal particles have been found to self-aggregate and exhibit nucleation centers of dendritic character. Earlier formed nuclei are more in number than those formed later and participated extensively in nucleation process.

The dendritic nucleus center has grown in a specific geometry, resulting in the DLA-branched crystal patterns during colloidal-phase reaction. Figure 2.1h presents the growth behavior and DLA patterns' branches. It propagates through side branching, which is similar to the symmetric dendrites. The patterns exhibited here is quite similar to the DLA patterns obtained experimentally in some other systems. Here, the growth initiates from prerequisite nucleation site like DLA model. The solvent medium induces the nanosized particles to attract and the aggregation occurs randomly towards nucleation center. Usually, nucleation happens with a particular concentration of monomer (ADA) over some time, but in the present case supply of monomer depletes with surface growth of clusters, resulting in the reduction of monomer concentration below the critical point; thus nucleating comes to the end. To understand the nanocrystal synthesis, knowledge on the growth process is very important. In smaller particles usually surface-to-volume ratio is quite high. In the case of small particles in a large surface area, excess energy on surface contributes considerably to the total energy. Thus, in a solution with non-thermodynamic equilibrium condition, larger particles form in the expense of smaller ones; reduction of the surface energy and growth of nanocrystals occur. The growth of the colloidal particles is initiated by diffusion of the sequence of monomer (ADA) towards the surface, and then these monomers react with each other at the surface of the nanocrystals, and totally depend on the surface energy. The interfacial energy is defined as the energy at an interface due to the differences between chemical potential value of the atoms at interfacial region and the atoms present in neighboring bulk phases. The chemical potential of a particle for a solid species at a solid/liquid interface usually increases with decrease in particle size, therefore smaller particles of a solute contains much higher concentration than the larger one. The equilibrium concentration of a nanosized particle present in liquid phase depends on the local curvature of the solid phase. There is least possibility for a fine solid particle to enter into the outer portion of a crystal. Regarding the screening effect of DLA-like experiment, nucleation centers formed at slower rate at the earlier stages in compared to those produced in later stage. These colloidal particles then aggregate to form DLA crystal patterns, which can be regarded as the growth units in the growth process.

2.3 Structural Determination of DLA Crystal Patterns

The morphological studies have been observed for the DLA crystal pattern with the help of optical microscopy, scanning electron microscopy, and transmission electron microscopy. The crystal patterns consist of a huge amount of crystals which are aligned properly. These branches are connected with a nucleus center as a common point, corresponding to the primary branches formed at the initial outline of the DLA-branched crystal patterns. Similarly, secondary and tertiary branches are in turn associated with primary branches resulting in the formation of densely branched crystal patterns. All these branches never get together during intersection due to diffusion field restrictions, implying that the aggregation process is not always random. The microphotographs of DLA-branched pattern are shown in Fig. 2.2.

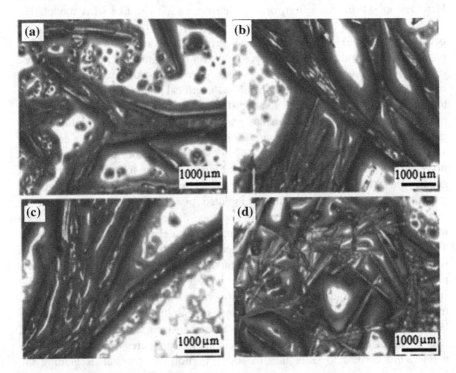

Fig. 2.2 OPM photograph of DLA-branched crystal pattern

This microphotograph appears with many centers and branches with long narrow wires bunched together, actually exhibiting a morphology of branch structure. In virtue of the periodic external perturbations, regular dendritic growth has been observed. The side branching is formed by expansions of growing tips, which becomes rough with newly formed branches, and predominate over others with progress in reaction.

Figure 2.3 represents the scanning electron microscopy and energy dispersive X-ray spectra of branched crystal pattern.

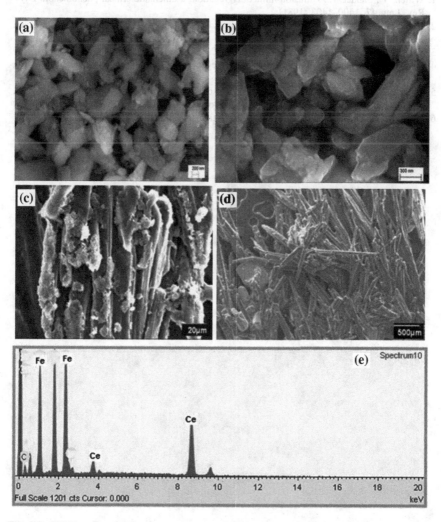

Fig. 2.3 SEM images **a** and **b** nanostructured crystalline particles **c** and **d** branched crystal pattern and **e** EDS spectrum

The TEM studies have confirmed the size of nanoparticles with ~20–100 nm diameters. From the TEM images it is clear that nanosized particles have self-aggregated in the colloidal phase as we have discussed in the Sect. 3.4, and finally they form the DLA crystal pattern via self-organization phenomenon

References

1. Witten, T., Sander, L.: Diffusion-limited aggregation, a kinematic critical phenomenon. Phys. Rev. Lett. **47**, 1400–1403 (1981)
2. Kobayashi, Y., Niitsu, T., Takahashi, K., Shimoida, S.: Mathematical modeling of metal leaves. Math. Mag. **76**, 295–298 (2003)
3. Batty, M.: Cities and Complexity. MIT Press, Cambridge, MA (2005)
4. Bourke, P.: Constrained limited diffusion aggregation in 3 dimensions. Comput. Graph. **30**, 646–649 (2006)
5. Halsey, T.: Diffusion-limited aggregation: a model for pattern formation. Phys. Today **53**, 36–47 (2000). http://www.physicstoday.org/pt/vol-53/iss-11/p36.html. Accessed Dec 2006
6. Voss, R.: Fractals in nature: from characterization to simulation. In: Peitgen, H., Saupe, D. (eds.) The Science of Fractal Images, pp. 36–38. Springer, New York (1988)

Chapter 3
Growth and Form of Spherulitic Crystal Pattern

3.1 Introduction

Mineral aggregates, simple organic liquids, and polymers [1–3] are common examples of spherulitic crystal pattern formation. The complex structure of spherulite is basically in the shape similar to needle-like crystal which is usually found in inorganic and organic materials, diverged from the center of the spherulite. The needle-like crystals are associated with the center of the spherulite to achieve a perfectly symmetrical shape.

The phenomena of crystal branching and spherulite formation are wide and diverse. This kind of morphogenesis is found in biological, geological (e.g., sediments), and synthetic systems as well [4]. Furthermore, the formation of spherulites has been reported for a large variety of materials [5]. The amazing architecture of natural and synthetic spherulites has attracted much interest although the mechanism of their formation is not yet definitely clear. One of the first studies on spherulites was published already in 1888 by Lehmann [6], who described them as radially arranged fibrillar aggregates. Since that time, a number of reviews and textbooks have been devoted to the description of spherulite formation and their properties. Most of these contributions are devoted to the understanding of crystal branching of "pure" compounds, crystallized according to so-called classical crystal growth mechanisms (crystal growth by ion-by-ion or molecule attachment to a primary particle).

3.2 Development of Spatial Pattern Formation

The development of spatial pattern formation in this type of BZ reaction could be observed in a two-dimensional Petri dish by pouring the reaction mixture consisting of ADA/Ce^{+4}/Fe^{+3}/BrO$_3$/H$_2$SO$_4$ system. I explored the utilization of dual substrate

© The Author(s) 2016
R. Srivastava et al., *Growth and Form of Self-organized Branched Crystal Pattern in Nonlinear Chemical System*, SpringerBriefs in Molecular Science, DOI 10.1007/978-981-10-0864-1_3

and dual catalyst which makes the system to give more dynamical phenomenon in comparison with normal BZ system, and the results of this observation could be more exciting and interested in the RD system. The initially homogenous red color parent solution rapidly turned into the blue with the construction of an interesting eye-catching wave pattern (Figs. 3.1a and 3.2). This phenomenon could be monitored within a couple of few seconds. After achieving a short steady state of period of time, the exiting color twisted into light red color with the construction of another interesting stationary-type pattern as represented in Fig. 3.1b. After its appearance, the pattern usually remains stationary for approximately 1 h before it progressively fades with maintaining two to three regular frequency oscillations, solutions bifurcates into blue color, where the reaction mixture reside for a couple of second. Again, after few minutes, the whole system jumped into a complex structure where the system bifurcates into another blue color stationary-type pattern (Fig. 3.1c). Gradually, it takes place in the form of colloidal material spread in the whole Petri dish where the system settles for a couple of hours and then bifurcates to non-homogenous solution for forthcoming several hours, and at the end, spherulitic branched crystal pattern forms are observed as shown in Fig. 3.1d. It was found that the observed spherulitic branched crystal pattern was stable for a couple of months. Now, the interesting point could be here what will be the probable mechanism of spatial pattern formation and growth of branched crystal pattern in the reaction diffusion system. Therefore, in this connection, we have proposed the mechanism of this phenomenon (Scheme 3.2). The fundamental advantages of this chemical system offer that the parameters can be easily control which determine the pattern formed with possible insights into the phenomenon close resemble to biological systems.

As the reaction starts and after 6 h, the colloidal state was observed in the solution phase and organizes in such a manner that numerous very fine solid particles spread throughout in the whole reaction solution. These particles aggregate with each other via self-assemble processes and form some discrete solid points, called solid-phase nucleation centers as is shown in Figs. 3.3 and 3.4 respectively which describes nucleation of solid-phase center, surrounded by homogenous dispersed colloidal particles. The SEM images (Figs. 3.7 and 3.8) confirm the colloidal phase of the reaction system.

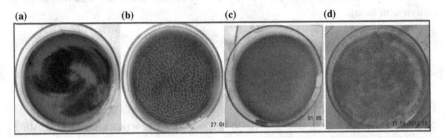

Fig. 3.1 Multiple pattern observed in ADA-AA-Ce^{+4}-$[Fe(Phen)_3]^{+2}$-BrO_3-H_2SO_4 system. The photograph was taken **a** min 5 s, **b** 2 min, and **c** 5 min after starting experiment [7]

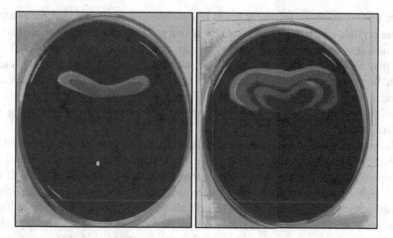

Fig. 3.2 Development of concentric ring in succinic acid (SA)-acetyl acetone (AA)-Ce^{+4}-[Fe (Phen)$_3$]$^{+2}$-BrO$_3$-H$_2$SO$_4$ system [8]

It was monitored that various fine particles initially aggregate with each other and form a coherent structure spread throughout in the whole Petri dish, and as a result, some new branched crystal patterns are formed. In the colloidal phase, the nucleation starts in the solid phase which was found to be slightly differed in character via spontaneously during complex reaction sequences. Before the nucleation processes, there might be possibility to form various types of intermediate products in the reaction mixture. The proposed reaction mechanism clearly suggests that these products significantly control the solid-phase nucleation in the reaction system. It has also been assumed that a concentration gradient exists between nucleation centers and uniformly distributed fine solid particles of the reaction mixture. The mechanism of this growth processes is helpful to explain the nucleation phenomenon as a diffusive mode. As we know, the excitability is the most important essential factor of the BZ-type reaction systems to show oscillatory phenomenon as well as pattern formation in the reaction diffusion system. Under excitable conditions, it has been observed that when the solid-phase nucleation processes start, the dynamic insta-bility and surface anisotropy, which are some non-equilibrium parameters, affect the nucleation growth phenomenon. Mullins–Sekerka type of dynamic instability is the best example to understand this phenomenon. In this experiment, this type of dynamic instability is usually found and concentration gradient generates between centers where the nucleation starts and symmetrically dispersed superior solid par-ticles. Therefore, this kind of instability often controls the mobility to combine together or to form crystal interfaces as it depends upon the diffusion coefficients of the colloidal particles. In another way, surface anisotropy determines the disorder in the growing crystal structure. It could be influenced by diverse form of crystals and also by a multicrystal phase. The formation of new nucleation center which exhibits some undeveloped crystal profile might be converted into some mesoscopic crystals. These crystals which are formed from the above processes have more surface energy and also less order crystallographic symmetry. The metal ion catalysts (Ce^{4+}/Fe^{2+}) can play an important role to break these stable faces to further in the direction of crystallographic axis, because they have both unsymmetrical crystal phases and sound clear crystallographic symmetry. Therefore, in such a way, the growth of orderly nucleation structure takes place in the whole direction by decreasing the surface energy.

In similar concept, the particle-mediated crystallization theory was used to explain well the formation of solid-phase nucleation in the reaction system. This theory suggested that some mesoscopic transformation process takes place in the solid phase followed by multiple nucleation growth phenomena. At the initial phase, it reorganizes into mesoscopic crystals followed by an orderly beautifully mesostructure via self-assembly, of well-aligned size and shape. The crystallographic orientation of the particle is equal in all directions so that the mesoscopic crystals can be reorganized in such a way under the processes of self-assembly. On the basis of the above explanation, Scheme 3.1 represents nucleation mechanism.

In Fig. 3.3, the growth processes of the spherulitic crystal patterns have been represented.

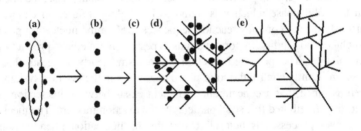

Scheme 3.1 Graphical representation of nucleation of solid phase

Fig. 3.3 Fully grown solid particles in the colloidal phase. **a** Respective SEM images of colloidal particles (**b**) [7]

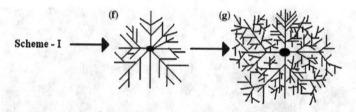

Scheme - I

(f)　　　(g)

Scheme 3.2 Schematic growth mechanism of spherulites

The characteristic morphology of the spherulite branched crystal pattern was found to closely resemble polymeric spherulites. Scheme 3.2 shows a schematic growth mechanism.

3.3 Structural Determination of Nanostructured Spherulite Crystal Pattern

The morphological structure of branched nanostructured spherulitic crystal pattern was characterized by with the help of optical microscopy, scanning electron microscopy, and transmission electron microscopy. It was found that the branched crystal pattern consists of more numbers of well-associated crystals. Each branching of the spherulitic crystal pattern associated with a nucleus center corresponds to the primary branches which form the initial outline of the spherulitic branched crystal patterns. Secondary and tertiary branches of the spherulite were in turn connected with primary branches which ultimately form the densely branched crystal patterns. It indicates that the aggregation process followed by the formation of branching crystal pattern was not being completely random. Figure 3.6 shows the optical microphotograph of the branched crystal pattern.

The microphotograph of spherulitic pattern represents various nodes and branching which looks like long thin wire bunched together. It is confirmed from the above images that morphology is princely branched. The side-branched positions of the spherulitic crystal pattern are apparent and interrelated on either side. The homogeneity of the structures is suggestive with regular dendritic growth subject to periodic external perturbations. In the figure, the side branching was found to be wide or expanded to a fixed distance. These growing tips intermittently become coarse by newly formed branches normally predominant over other as reaction time progresses (Fig. 3.5).

The morphology of the spherulitic branched crystal has been investigated by SEM technique. The SEM images of the spherulite-like branched pattern are shown in Figs. 3.7 and 3.8. In the SEM images, multiple nucleation centers have been occurred which self-regulate to each other in the form of the spherulitic branched

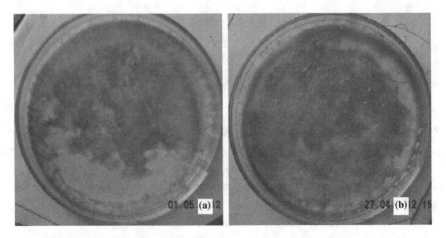

Fig. 3.4 Growth of spherulite. **a** After 3 h and 40 min and **b** after 6 h and 20 min

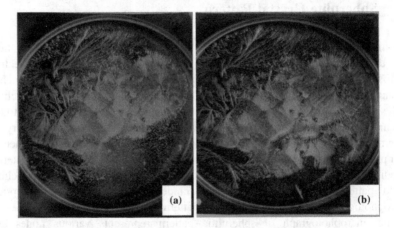

Fig. 3.5 Growth of spherulitic crystal pattern in SA-AA-Ce^{+4}-[Fe(Phen)$_3$]$^{+2}$-BrO$_3$-H$_2$SO$_4$ system. **a** After 13 h 15 min and **b** after 13 h 45 min [8]

crystal patterns of different morphology. The SEM images of the branched crystal pattern clearly shows the shape of the nanostructured particle are cubic in nature, and some of them hexagonal. Diverging crystal fibrils are also emphasized and provide evidence of birefringency or Maltesecross extinction patterns similar to polymer spherulites. The SEM images of the spherulitic crystal pattern also represent a number of various fine solid crystal fibrils organized in the form of bundles. Some crystal aggregates are also formed. It has been suggested that these might be the residue of metal ion catalyst or non-crystalizing amorphous material. These materials could facilitate the binding together with the multicrystal phase and also play a role in modulating the spherulitic frame during the self-assembly processes (Figs. 3.7 and 3.8).

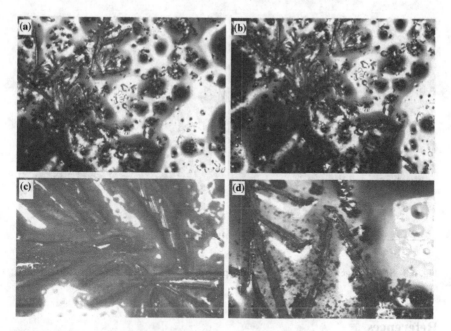

Fig. 3.6 OPM images of spherulitic branched crystal pattern [7]

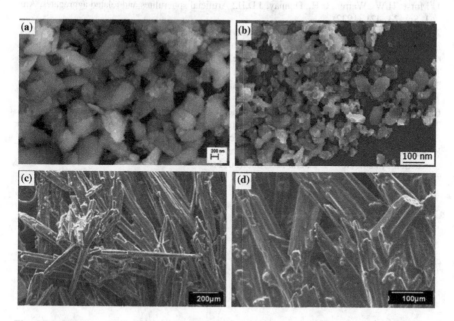

Fig. 3.7 SEM images of nanostructure crystalline particles. **a** and **b** SEM images of spherulitic crystal pattern **c** and **d** in ADA-AA-Ce^{+4}-[Fe(Phen)$_3$]$^{+2}$-BrO$_3$-H$_2$SO$_4$ system [7]

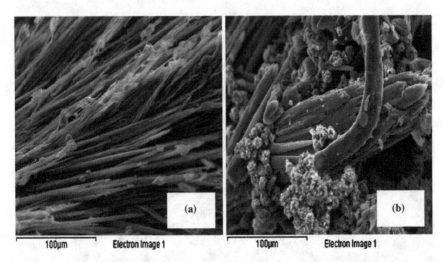

Fig. 3.8 SEM images of spherulitic crystal pattern SA-AA-Ce^{+4}-[Fe(Phen)$_3$]$^{+2}$-BrO$_3$-H$_2$SO$_4$ system [8]

References

1. Morse, H.W., Warren, C.H., Donnay, J.D.H.: Artificial spherulites and related aggregates, Am. J. Sci. **23**, 421 (1932)
2. Sperling, L.H.: Introduction to Physical Polymer Science, Chap. 6. Wiley, New York (1992)
3. Hutter, J.L., Bechhoefer, J.: J. Cryst. Growth **217**, 332 (2000)
4. Prymak, O., Sokolova, V., Peitsch, T., Epple, M.: he crystallization of fluoroapatite dumbbells from supersaturated aqueous solution. Cryst. Growth Des. **6**, 498 (2006)
5. Punin, O., Shtukenberg, A.G.: Autodeformation Defects in Crystals. St. Petersburg University Press, St. Petersburg (2008). (in Russian)
6. Lehmann, O.: Molecularphysik, vol. 1. Wilhelm Gugle-mann, Leipzig (1888)
7. Srivastava, R., Srivastava, P.K.: Self-organized nanostructured spherulitic crystal pattern formation in Belousov-Zhatobinsky type reaction system. Chem. Phys. **426**, 59–73 (2013)
8. Yadav, N., Srivastava, P.K.: Growth of spherulitic crystal pattern in Belousov-Zhabotinski type reaction system. New. J. Chem. **35**, 1080 (2011)

Chapter 4
Summary of the Research Work

In this book, the new different types of modified BZ reaction system have been investigated with a view to understand the mechanism of spatial pattern formation and synchronize self-organized formation of ordered nanostructured dendritic structures (DLA and spherulitic crystal structure). The formation of dendrite structure based on the above processes is an important example of morphogenesis in Laplacian field. The core of the research is that a nanostructured DLA and spherulitic crystal pattern formation phenomena by utilizing a chemical system to control nanostructures. Pattern formation in these types of modified BZ systems involving dual substrate and dual catalyst has been optimized in order to understand that what are the important parameters which could control this kind of spatial pattern formed. The BZ system involving dual substrate and dual catalyst and microemulsion system have been investigated for the first time. It has been investigated that the formation of nanostructured DLA and spherulitic crystal pattern takes place only when there is a formation of intermediate spatial pattern in solution phase. It was observed that during the formation of such structures in BZ system involving dual substrate and dual catalyst, UV–visible spectroscopy could be effectively used to probe the intermediate stages leading to such phase transformation. It was found that the amplitude of the absorbed peak, corresponding to the coexisting ferroin in the reaction system, started oscillating typically after a certain interval of time from the mixing processes of all the BZ reactants. It was also found that oscillatory behavior exhibited by UV–visible spectroscopy and the growth of nanostructured DLA and spherulitic crystal patterns are interrelated. In addition, the description of the general mechanism has promoted us to categorize the oscillatory dendrite formation into three different systems according to the differences in the size and shape of the crystals formed.

The general mechanism and the classification of the present work in this book have provided a promising and interesting way to prepare, design, logically organize, and control the nanostructured materials during the modified BZ reaction involving dual substrate and dual catalyst. The study is believed to be a new addition in the field of nanoscience and can help the research scientific community

© The Author(s) 2016
R. Srivastava et al., *Growth and Form of Self-organized Branched Crystal
Pattern in Nonlinear Chemical System*, SpringerBriefs in Molecular Science,
DOI 10.1007/978-981-10-0864-1_4

acquire some insight into the pattern formation mechanism which is close relevant to the pattern formation in biological, physical, and chemical complex processes which also occur in our nature.

The outline of the research work highlighted in this book opens new possibilities for future research as different ranges in the parameter space remain unexplored and unexpected dynamics may appear.

Chapter 5
Future Prospects

So far, in this book, we have investigated the growth of nanostructured dendrite structure in BZ-type reaction involving dual substrate and dual catalyst and oscillatory behavior could be probed by spectrophotometric methods. Now, there is a need to study the chemical oscillation in terms of other parameters. It will be an interesting task to investigate how the nanostructured DLA and spherulitic crystal patterns can be controlled and designed by varying some parameters, e.g., the temperature, concentration of individual BZ reactant, thickness of the solution, and by varying the stirring rate. The spatial pattern formation has been well seen in systems we have explored. Further, it will be interesting to search whether by using dual substrate and dual catalyst or microemulsion system one could materialize the formation of turing pattern and spiral wave. The nearest future task will be to analyze the crystallographic structures of the as-grown nanostructured DLA and spherulitic crystal patterns. In microemulsion system, the spatial pattern formation followed by nanostructured DLA pattern has been investigated. There is now a need to study the oscillatory behavior in terms of governing parameters and try to correlate with the formed dendrite structure. Further, it will be important and necessary to analyze the effect of the period and amplitude of the BZ reaction on DLA crystal pattern formation and morphological transition. More importantly, the scientific mechanism behind the phenomena of nanostructured DLA crystal pattern and the relationship between nanostructured DLA pattern formation and BZ-type reaction could be revealed. The various spatiotemporal pattern and formation of fractal pattern could also be investigated in o-hydroxyacetophenone—Ce^{4+}–BrO_3^-–H_2SO_4 system dual frequency oscillator. Further study will be to establish the mechanism of the investigated chaotic oscillation in this BZ type sequential oscillator. The research work carried out here is totally based on the experimental observation; it will be further interesting to design and produce a mathematical model/numerical simulation for the growth of these DLA and spherulitic crystal patterns.

In particular, another attempt of the research in the field of nonlinear chemical dynamics could be to produce and to study "exotic phenomena" in chemical systems, to reveal their mechanism, and to simulate these temporal and spatial patterns.

© The Author(s) 2016

R. Srivastava et al., *Growth and Form of Self-organized Branched Crystal Pattern in Nonlinear Chemical System*, SpringerBriefs in Molecular Science, DOI 10.1007/978-981-10-0864-1_5

Oscillatory chemical reactions are possibly the simplest systems in which nonlinear dynamics can be studied. The conclusions drawn from the studies at the molecular level (i.e., in chemical systems) will hopefully lead to some general rules and help us to understand similar phenomena observed in more complex systems; in biology, technological processes or in the prediction of periodically occurring crises in society, etc.

About the Book

The book introduces the oscillatory reaction and pattern formation in the Belousov–Zhabotinsky (BZ) reaction that became model for investigating a wide range of intriguing pattern formations in chemical systems. So many modifications in classic version of BZ reaction have been carried out in various experimental conditions that demonstrate rich varieties of temporal oscillations and spatiotemporal patterns in non-equilibrium conditions. Mixed-mode versions of BZ reactions, which comprise a pair of organic substrates or dual-metal catalysts, have displayed very complex oscillating behaviors and novel space–time patterns during reaction processes. These characteristic spatiotemporal properties of BZ reactions have attracted increasing attention of the scientific community in recent years because of its comparable periodic structures in electrochemical systems, polymerization processes, and non-equilibrium crystallization phenomena. Instead, non-equilibrium crystallization phenomena which lead to the development of novel crystal morphologies in constraint of thermodynamic equilibrium conditions have been investigated and are said to be stationary periodic structures. Efforts have continued to analyze insight mechanisms and roles of reaction–diffusion mechanism and self-organization in the growth of such periodic crystal patterns. In this book, non-equilibrium crystallization phenomena, leading to growth of some novel crystal patterns in dual organic substrate modes of oscillatory BZ reactions, have been discussed. Efforts have been made to find out experimental parameters where transitions of the spherulitic crystal patterns take place. The book provides the scientific community and entrepreneurs with a thorough understanding and knowledge of the growth and form of branched crystal pattern in reaction–diffusion system and their morphological transition.

© The Author(s) 2016
R. Srivastava et al., *Growth and Form of Self-organized Branched Crystal Pattern in Nonlinear Chemical System*, SpringerBriefs in Molecular Science,
DOI 10.1007/978-981-10-0864-1

Printed in the United States
By Bookmasters